New Standards for Fans

Organizing Member

W T W Cory
Woods Air Movement Limited, UK

IMechE
Seminar Publication

I MECH E

New Standards for Fans

3 October 2000
IMechE Headquarters, London, UK

Organized by
The Fluid Machinery Committee of
The Power Industries Division of
The Institution of Mechanical Engineers (IMechE)

IMechE Seminar Publication 2000–14

**Professional
Engineering
Publishing**

Published by Professional Engineering Publishing Limited for The Institution of
Mechanical Engineers, Bury St Edmunds and London, UK.

First Published 2000

ISSN 1357–9193
ISBN 1 86058 311 3

A CIP catalogue record for this book is available from the British Library.

Printed by The Cromwell Press, Trowbridge, Wiltshire, UK.

Contents

Related Titles of Interest

Title	Editor/Author	ISBN
Process Fan and Compressor Selection	J Davidson and O von Bertele	0 85298 825 7
Developments in Turbomachinery Design	J Denton	1 86058 237 0
IMechE Engineers' Data Book – Second Edition	C Matthews	1 86058 248 6
Handbook of Mechanical Works Inspection – A guide to effective practice	C Matthews	1 86058 047 5
Improving Maintainability and Reliability through Design	G Thompson	1 86058 135 8
Process Machinery – Safety and Reliability	W Wong	1 86058 046 7
CFD in Fluid Machinery Design	IMechE Seminar	1 86058 165 X
Installation Effects in Fan Systems	IMechE Seminar	1 86058 107 2
Noise in Fluid Machinery	IMechE Seminar	1 86058 246 X

For the full range of titles published by Professional Engineering Publishing contact:

Sales Department
Professional Engineering Publishing Limited
Northgate Avenue
Bury St Edmunds
Suffolk
IP32 6BW
UK

Tel: +44 (0)1284 724384
Fax: +44 (0)1284 718692

An introduction to the need for new standards for fans

W T W CORY
Woods Air Movement Limited, Colchester, UK

1. INTRODUCTION

Until the early 1920s, the methods for testing the aerodynamic performance of industrial fans were legion. It is no exaggeration to say that these were determined by the various manufacturers according to their own beliefs, prejudices or downright commercial considerations. At that time ASHVE in the USA, and IHVE in the United Kingdom both set up fan standardisation committees which produced recommendations for the conduct and calculation methods to be used in such tests. Subsequently these recommendations were incorporated within National Standards, but the proliferation continued. It seemed that we had simply exchanged one set of problems for another, as ever more organisations felt qualified to issue a standard. Not only did ASME issue their own standard in the USA, but the fan Manufacturers Association in the UK also issued its own code. Other National bodies joined the game so that by the year 1978 it was estimated that there were at least eleven national fan performance test codes with 65 distinct test methods in use.

Into the chaotic situation which existed, ISO stepped with great confidence. It set up technical committee TC117 in 1963 to discuss the formulation of an International Standard which could be agreed by all the industrial fan manufacturing nations. It started off with considerable optimism and after various excursions along the way - performance to be quoted as mass flowrate kg/s against specific energy J/kg - eventually settled into a dull routine where each nation sought to protect its own code, at the expense of the others. Eventually it dawned that compromise was essential if we were to complete our work this side of the grave.

You may well ask - Why all the fuss? Does it come as a surprise to know that not all those 65 distinct methods were of the same technical merit and serious discrepancies could result? A few years ago my company carried out tests on one particular fan to various codes and the **supposed** differences in performance **Figure 1** are alarming. In fact, of course, nothing should have changed. If efficiencies had been plotted, with an unchanged fan power, then at a given flowrate these would be proportional to the fan pressure.

It should be noted that the fan was a Tube Axial with appreciable outlet swirl. How this swirl energy is treated can have an appreciable effect on the results and is the subject of Mr M. Stevens paper.

This is a world which endeavours to preach the value of free trade. Increasingly it will have to accept the fact of globalisation. As its contribution to harmony between nations, it is a little strange that valid comparisons could not be made.

2. THE MEASUREMENT OF AIRFLOW

The fan engineer works under an extreme disadvantage. His creations handle a fluid which cannot be seen or directly weighed. If necessary a pump flow could be determined by catching the water in a bucket. We don't have the possibility of determining airflows in that way.

Furthermore, in our real world, air travels in three dimensions and is turbulent. If one is making measurements under actual installation conditions it is therefore desirable to take a great many measurements of velocity and direction. This is the philosophy behind ANSI/ASME PTC 11-1984 which is a code developed in the USA for determining performance under operating conditions. Typically it is used for the acceptance tests of large fan units such as those required for Mechanical Draught in central power stations. It normally requires the use of a calibrated 5 hole pitôt tube **Figure 2** combined with a temperature sensor. A traverse is taken directly on the fan discharge and the many measurements of pressure (total and static), direction (pitch and yaw) and temperature (wet and dry-bulb) are then integrated to obtain the total flow and pressure. This necessitates the use of a computer for the otherwise tedious hand calculations.

The alternative to this is to collect, steady and generally organise the flow in a suitable test airway, and this is the aim of the various laboratory test methods. In this way a number of devices can be used which enable the flow to be calculated from a pressure difference. These are less susceptible to operator skill level than the pitôt tubes previously favoured.

3. FAN PERFORMANCE

The performance of a fan is affected by the connections made to its inlet and outlet. Ducting, where fitted, not only has a pressure loss, but can act as an impedance, modifying the flow into or out of the fan casing. In extreme cases it can even prevent the development of a full velocity profile. Ideally the flow velocity sectors should be symmetrical and axially aligned (free from yaw) and without swirl or spin (pre or contra) if the fan is to develop its design duty.

From the infinite variety of possible connections, two are normally selected as representative of the range of performance:

a)   The fan inlet or outlet is open directly to the unobstructed free atmosphere;
b)   The fan inlet or outlet is open directly to a long, straight duct of the **same area** as the inlet or outlet.

What, therefore, have been the major resolutions of ISOTC117, and which have been incorporated in ISO5801?

(i)   It recognises that a fan will perform differently according to how it is installed.
Type A with free inlet and outlet
Type B with free inlet and ducted outlet
Type C with ducted inlet and free outlet
Type D with ducted inlet and outlet.

It will be seen that the two alternative connections previously mentioned have been combined to give the four possible installation categories **Figure 3**. In installations of type A, a partition in which the fan is mounted may support a pressure difference between the inlet and outlet sides.

(ii)   It introduces the concept of 'common parts' of the ducting adjacent to the fan inlet and/or outlet sufficient to ensure and accurate and consistent determination of fan pressure no matter what method of flow measurement or control is used. The dimensions of these parts have been specified such that the duct area must be closely matched to the fan inlet/outlet area as relevant, whilst their length is generally longer than those previously used **Figure 4**.

(iii)   It specifies the use of a 'conditioner' on the outlet of type B or D fans. This is designed to dissipate any swirl energy which is not normally available for overcoming the system resistance. *It has been noted that the design of duct systems is based on the addition of pressure losses in the various elements. These losses are taken from text books which all assume a fully developed symmetrical and swirl free flow both upstream and downstream of the element. The losses in the presence of swirl would be appreciably higher.*

(iv)   When a fan is tested with a conical inlet for flow measurement, the declared results using the new standard may be lower as the values of discharge co-efficient have been reduced when compared with some National Standards. Fan pressure is now defined as the difference in **stagnation** pressure between the fan inlet and outlet.

4. INSTALLATION CATEGORY

The differences in fan performance according to installation category are as much a function of the fan type and design, as of the position of the duty point on the particular characteristic curve.

In practice the type B and type D characteristics for most fan types will be nearly the same at the best efficiency point, provided the fan is supplied, in its free inlet form, with a properly shaped entry cone or bellmouth. With the same proviso, type A performance will coincide with type C **Figures 5 & 6**. The essential difference remaining is that between free outlet and ducted outlet performance, which is significant for fans of all kinds, though it diminishes as the ratio of fan velocity pressure to fan total pressure falls, and will be affected by tongue pieces in a centrifugal fan outlet. In the latter, a length of ducting is desirable to enable some recovery of dynamic pressure to useful static pressure to be achieved from the distorted velocity profile.

It should be noted that the claim made in many fan catalogues that the absence of an outlet duct is equivalent to a performance loss of two fan velocity pressures as calculated from the discharge area is a very crude approximation. This can only be correct for one particular duty point and would require a **very** flat, very symmetrical and swirl free velocity profile at the discharge.

5. COMMON PARTS OF DUCTING

Satisfactory measurement of pressure cannot be taken immediately adjacent to the fan inlet or outlet and it is necessary to establish test stations some distance away, where the flow can be normalised.

Tests at CETIAT (France) had shown how oversized ducts had 'enhanced' fan performance, whilst NEL (Britain) had found that insufficient length could also result in inaccurate measurement of fan pressures. Both sets of results were used when fixing dimensions.

(a) The common parts include a duct on the outlet side of the fan, having a length of five equivalent diameters to the pressure measuring point and incorporating a standardised flow straightener. Without such parts, differing values of pressure can result according to the character of the airflow at the fan outlet.

The velocity distribution at this point often contains considerable swirl. Even when free from swirl it is far from uniform. This results in an excess of kinetic energy or velocity pressure over the conventional allowance of $\frac{1}{2}\rho v^2$, caused by the proportionality of kinetic energy to the local value of ρv^3 (mass flow x velocity pressure) so that the excess where v is high exceeds the deficit where v is low.

Now the non-uniformity of **axial** velocity components diminishes as the flow proceeds down the duct and the excess energy reaches a minimum of a few percent of $\frac{1}{2}\rho v^2$ within a length equal to two or three duct diameters. Part of the original excess is lost, but part is converted into additional static pressure, the conventional velocity pressure remaining constant. This addition to the fan pressure is available for overcoming external resistance, and in order to credit it to the fan, as it should be for type B and type D installations, it has been determined that the test station for outlet side pressure measurement should therefore be more than three duct diameters from the outlet **Figure 7**. In fact, to accommodate the étoile flow straightener, it has been set at five diameters as previously noted.

a) A transition section may be used to accommodate a difference of area and/or shape but to minimise the effects of any change in aerodynamic impedance, it is specified that the duct area shall be within the limits of 5% less or 7% more than the fan discharge area. The dimensions of the transition are also specified to give a small valley angle.

b) Common parts on the fan inlet are shorter and the pressure measurement station need be only three equivalent duct diameters from the fan inlet. This reflects the more regularised conditions which apply on this side. For the same reasons, in an accelerating flow, a greater deviation in the upper limit of duct diameter is permitted. The lower limit is set at 5% less areas of duct to fan inlet. Again the transition angles are specified to minimise the effects of flow separation.

6. FLOW CONDITIONERS

The **swirl** energy at the fan outlet is rarely recovered in a straight uniform duct and then only over very long distances – more than 100 diameters. In the presence of swirl simple measurements of effective pressure or volume flow rate are impossible, and it must therefore be removed when tests are to be taken in a duct on the outlet side of the fan, to give information on type B or D performance. An effective flow straightener or conditioner will do this. If it removed just the swirl energy and no more, the minimum energy convention would be satisfied. However, the energy actually removed is very dependent on the combination of swirl pattern and straightener. Again, the need for an agreed standard outlet duct will be appreciated.

In practice a fan with a lot of outlet swirl ought not to be selected for use with a **long** straight outlet side duct, because the friction loss in the latter will be substantially increased as the 'wetted' path will be longer. Guide vanes should be fitted which will remove and recover (instead of removing and destroying) the swirl energy. The flow straightener will then just ensure that test conditions are satisfactory in the downstream duct: the relatively small outlet swirl components from centrifugal, guide-vane axial or contra-rotating fans will be removed without measurable disturbance to the performance.

The actual design of straighteners to be used in the standardised test ducts is therefore of great importance. It is appropriate to review the two types which were incorporated, and which are also used in ISO Standard 7194 Measurement of fluid flow in swirling or asymmetric conditions.

(a) The AMCA straightener **Figure 8a** is used only to prevent the growth of swirl in a normally axial flow. It does not improve asymmetric velocity distributions. The device consists of a nest of equal cells of square cross-section. It has a very low pressure loss and is typically used either side of an auxiliary booster fan where this is necessary to overcome the resistance of the airway when a complete fan characteristic is required.

(b) The Etoile straightener **Figure 8b** is again designed to eliminate swirl but is of little use in the equalisation of asymmetric velocity distributions. The eight radial vanes should be of sufficient thickness to provide adequate strength but should be of sufficient thickness to provide adequate strength but should not exceed $0.007 \, D_4$ for pressure loss considerations. This straightener has a similar pressure drop to the AMCA straightener i.e. approximately 0.25 times the approach velocity pressure, but is also easier to manufacture. More importantly, it allows the static pressure to equalise radially as the air flows through it. This is not the case with the AMCA straightener which can produce variations in the static pressure across the duct downstream. The étoile straightener is therefore **preferred** in the common duct on the fan outlet.

It should be noted that well designed centrifugal fans or axial and mixed flow fans with efficient outlet guide vanes will not be penalised at design duty by the incorporation of flow conditioners in the test ducting. However, an axial flow fan without outlet guide vanes could be penalised by the ISO standard up to as much as 13 points on peak efficiency and over 20% on pressure. Centrifugal fans with poor outlet velocity profiles may also suffer. When operating away from the best efficiency point i.e. 'off-design', residual swirl may be present in all types of axially ducted fans, such that the straightener will reduce the pressure developed **Figure 9**.

7. DISCHARGE COEFFICIENTS

The standard will permit the use of a number of different types of primary flow measurement devices such as Venturi nozzles, orifice plates or conical inlets. Whilst the Venturi is a preferred device because of its relatively low pressure loss and insensitivity to disturbances in the approaching airflow, it has to be recognised that it can be expensive, especially when a range of sizes are required. Orifice plates incur higher pressure losses and booster fans may have to be inserted in the test arrangement to enable the fan characteristic to be measured down to zero pressure and maximum flow. Inlet cones are therefore of great importance.

The general expression for the mass flow rate through all these devices may be obtained from

$$q_m = \alpha \; \varepsilon \frac{\pi d^2}{4} \sqrt{2\rho u \, \Delta\rho}$$

where q_m is the mass flow rate   kg/s
       d is the throat diameter of the device   m
       ρu is the upstream air density   kg/m^3
       $\Delta\rho$ is the pressure difference across the device   Pa
          α is the flow coefficient
          ε is the expansibilty factor

α and ε may be combined to produce a compound flow or discharge coefficient C_D. This is generally plotted as a function of the local Reynolds number. Recent work on various International Standards has suggested that all these devices may be used with equal confidence provided that their discharge coefficients are based on the latest experimental data. For the conical inlet it has been noted that the coefficient seems to be related to boundary layer effects. Larger inlets have higher coefficients then those for small inlets.

8. THE MEASUREMENT OF FAN PRESSURE

One of the major decisions of TC117 was that the Fan Pressure should be defined as the difference between the stagnation pressure at the fan outlet and the stagnation pressure at the fan inlet.

Stagnation pressure p_{sg} should not be confused with static pressure p_s. In fact up to about 2.5kPa it is virtually the same as the previously specified **total** pressure p_t.

The absolute stagnation pressure at a point p_{sg} is the absolute pressure which would be measured at a the point in a flowing gas if it were brought to rest isentropically i.e.

$$p_{sg} = p\left(1 + \frac{\kappa - 1}{2}Ma^2\right)^{\frac{\kappa}{\kappa-1}}$$

where   p   is the absolute pressure

         κ is the insentropic coefficient $= c_p/c_v$

         Ma   is the Mach number

9. THE EVENTUAL ISO5801

In 1997 after come 34 years of effort, ISO5801 was published. We now have an internationally accepted Fan Test Standard. It is suggested that this should be the basis of all specifications where comparisons have to be made between internationally competing products. It has to be said that even now it is not perfect. Some relaxations have been made to the requirements for the 'common part' on the outlet side of the fan.

The methods of calculation are overly complicated. There are in fact four approaches according to the degree of compression and also the Mach number of the air. When the discussions started, computers were not generally available. I suppose that we all wanted the simple approach and only accepted complexity when it was absolutely necessary. We were in fact slide-rule literate rather than computer literate. Now that PCs are invariably used, it may be that we should only accept the exact formulae which take into account all the effects of compressibility, Mach number, change of datum etc.

Nevertheless, it is a great stride forward. The UK and Italy have each adopted it as their National Standard without change, and it is understood that other countries will be following suit, dual numbering it as necessary.

10. SITE TESTING TO ISO5802

Perhaps it is only natural that this introductory paper should have devoted so much space to TC117's magnum opus - ISO5801. However, a site test standard, constructed along similar lines was deemed an inevitable 'follow-on'. Work on ISO5802 has now been completed and this should be published within the year. It adopts much of the philosophy of ISO5801 i.e., the definitions of the fan pressure, how to measure fan flow etc. It is obvious however that ducting common parts may not exist and the position of measuring stations has to be defined to ensure acceptable velocity profiles. Uncertainties of results under site conditions also have to be recognised and quantified. System effect factors however, are **not** quantified (perhaps these must await a future standard) but certainly their existence is recognised.

11. OTHER FAN STANDARDS

Flushed with all this success, ISOTC117 has made advances in a number of other areas. There are many qualities associated with a fan which have to be specified, apart from its aerodynamic performance. These are the subject of other papers in this volume. Topics covered include noise, vibration and balance, dimensions, performance presentation and tolerances, terminology etc. A method for measuring the thrust of jet tunnel fans has also been completed.

For reference purposes, it is felt that a table of these standards will be of use. All these ISO documents will be adopted by BSI, unaltered and in their entirety, as national standards, dual numbered as shown.

ISO Standard	Title	BS Number
5801	Industrial fans - Performance testing using standardised airways	848 Part 1
† 5802	Industrial fans - Performance testing in-situ	848 Part 3
12499	Industrial fans - Mechanical safety of fans - Guarding	848 Part 5
† 13347 - 1	Industrial fans - Methods of noise testing - General overviews	848 Part 2.1
† 13347 - 2	Industrial fans - Methods of noise testing - Reverberent room methods	848 Part 2.2
† 13347 - 3	Industrial fans - Methods of noise testing - Enveloping surface methods	848 Part 2.3
† 13347 - 4	Industrial fans - Methods of noise testing - Sound intensity methods	848 Part 2.4
† 5136	Acoustics - Determination of sound power radiated into a duct by fans and other air moving devices - Induct method.	848 Part 2.5
10302	Acoustics - Method for the measurement of airborne noise emitted by small air-moving devices.	848 Part 2.6
† 13348	Industrial Fans - Tolerances, Methods of conversion and technical data presentation.	848 Part 9
13349	Industrial Fans - Vocabulary and definitions of categories	848 Part 8
13350	Industrial Fans - Performance testing of jet fans	848 Part 10
13351	Industrial Fans - Dimensions	848 Part 4
† 14694	Industrial Fans - Specification for balance quality and vibration levels.	848 Part 7
† 14695	Industrial Fans - Method of measurement of fan vibration.	848 Part 6

† In preparation

And that is not all. Within the European Union we are obliged to comply with Directives. Those of most importance to the fan community are to do with :

i)   Machinery (Safety)
ii)   Low voltage
iii)   Electro-magnetic Compatability
iv)   Atex (explosive atmospheres)

It is inevitable that CEN 'C' type machinery specific standards will be written to cover these requirements. Professor Hans Witt in his papers will describe progress on fan standards for Machinery Safety and Explosive atmospheres.

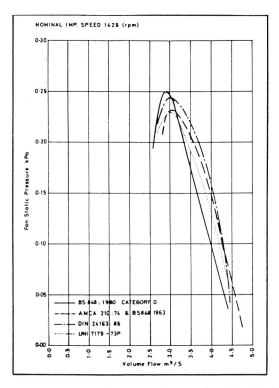

Fig. 1. Comparison of tests to various national standards

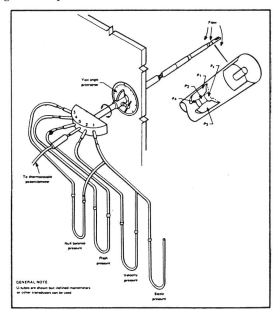

Fig. 2 5 hole pitôt tube

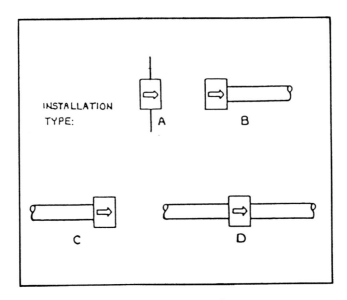

Fig. 3 Standard fan installation types

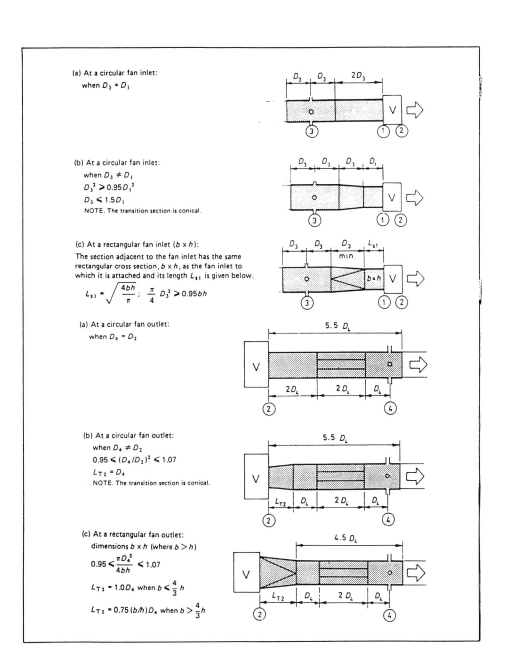

Fig. 4 Common parts for ducting on fan inlet and outlet

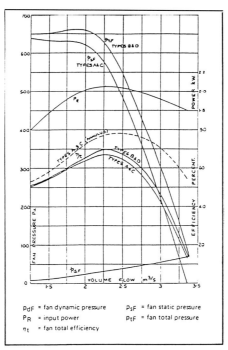

Pₐғ = fan dynamic pressure   Pₛғ = fan static pressure
P_R = input power   Pₜғ = fan total pressure
ηₜ = fan total efficiency

Fig. 5 Performances of a mixed flow fan in installations of type A, B, C and D

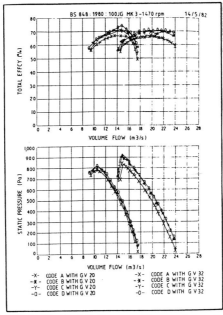

Fig. 6 Effect of test installation on guide vane fan performance

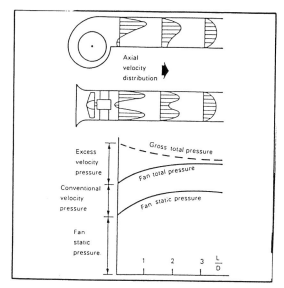

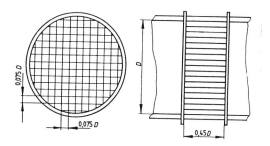

Fig. 7 Velocity diffusion downstream of a fan

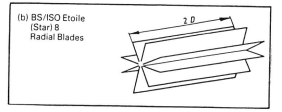

(a) AMCA Multi-cell

(b) BS/ISO Etoile
    (Star) 8
    Radial Blades

Fig. 8 Standard flow straighteners or conditioners

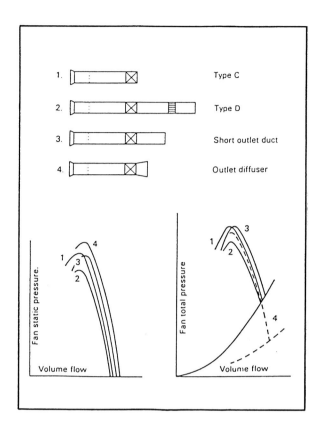

Fig. 9   Fan characteristic with outlet swirl

© 2000 With Author

S723/002/2000

The CEN standard for fans – mechanical safety

H G D WITT
Witt and Sohn, Pinneberg, Germany

SYNOPSIS

A CEN committee has produced a draft EN standard concerning the mechanical safety of fans, which is presented here. The standard is intended to specify in some detail the requirements of the European Machinery Directive for the component "fan". The paper contains a list of hazards specific to fans and of safety requirements and measures dealing with these hazards. It also contains chapters dealing with the proper way of fan markings and users instructions.

1. SCOPE

The European Standard will be applicable to all types of industrial fans. It does not apply to fans intended for household or similar purposes. The standard lists the significant hazards associated with fans and specifies safety requirements applicable to the design, installation, operation, maintenance and dismantling of fans during their foreseeable lifetime and subsequent disposal.

This standard is not exclusive and where specific standards exist relating to special applications these must be followed.

Low power fans intended for use on a single phase AC at a voltage not exceeding 250 V and a current not exceeding 16 A may alternatively be covered by the requirements given in CEI/IEC 60335-2-80. For explosion proof fans a separate specific EN standard is presently being developed.

2. FAN SPECIFIC HAZARDS

There are many fan specific hazards. Besides the obvious mechanical and electric hazards and the hazard that fans may cause ignition of combustible material, which is excluded from the scope of the standard, there are:

Thermal hazards
Noise hazards
Hazards due to vibration
Hazards due to the medium handled by the fan
Hazards generated by neglecting ergonomic principles in machine design
Hazards caused by failure of the energy supply
Hazards caused by missing or incorrectly positioned safety measures
Hazards due to transport, erection or bad maintenance

Some of these risks may have little practical importance. Only the required counter measures for the more common risks it will be mentioned here. For details the standard should be consulted.

3.1. Mechanical Safety

Primarily guards shall be provided in order to prevent contact with moving parts. Guards as well as inspection hatches must be fixed by systems that can be opened only with tools. If safety is to be ensured by the plant user he must be notified correspondingly, either by a notice on the fan or in the instructions for use. The fan proper must be safe against ejection of parts, bursting or breakage. The potential risks from auxiliary components as pressurised equipment must be eliminated as well.

3.2. Electrical Safety

All electrical components of fans must comply with the corresponding safety rules. Protective devices shall be employed to make the fan "fail safe". The build-up of electrostatic charging shall be avoided by proper earthing.

3.3. Thermal Safety

Hazards to personal due to thermal influences, which can arise from permanently high or low local temperature levels relative to the surrounding or from sudden fluctuations in temperature, e.g. as a result of a fire or an explosion, shall be kept as low as possible.

To this end it may be required to guard or insulate fans or components having temperatures above 70° C or below minus 10° C. A warning may also be sufficient. Nothing specific is said about electric motors, which generally have a higher temperature than 70° C. It would seem that a warning in the users instruction would be required and sufficient.

3.4. Noise

Design considerations shall be given to the noise emission values. Specific levels are not defined.

3.5. Vibrations

In order to avoid excessive and potentially destructive vibrations fans must be properly balanced according to existing standards. Deterioration of impeller balance due to encrusting or abrasion in conveying fans must be taken into account. If required proper operational methods must be adopted as monitoring of vibrations or cleaning devices. In case of clogging components in the air steam also design measures may be required, e.g. a special design of impeller blades.

3.6. Materials and Substances

The escape of hot or noxious gases must be prevented. Only proper constructional materials shall be employed. If certain gases are used in sealing, these shall not be harmful.

Hazards arising from the formation of mould and other micro-organisms in the fan shall be avoided. Drain outlets shall be provided as well as cleaning access to the interior of the fan.

3.7. Ergonomic Principles

Instruments, control devices, greasing points etc shall be located so as to be easily visible or accessible from the operators position.

Where relevant the fan shall be designed to give safe access during purging, draining, etc.

3.8. Failure of Energy Supply

In case of loss of the main - or auxiliary energy supply it must be possible to bring the fan to a safe condition. When the energy is reinstated that must not lead to a dangerous situation.

Special care is required if leakage of noxious gases may increase at standstill or if the cooling of hot gas fans is impeded.

3.9. Failure of Fan Parts

All parts and components shall be suitable for the intended use, in particular as regards durability, fatigue, aging, unbrittlement, corrosion, abrasion, chemical reactions, heat, electrostatic effects and sparking. They shall also be suitable for the agreed temperature limits. Care shall be taken that frequent speed variations will not lead to material fatigue.

Transient pressure resistant fans, gas tight fans, abrasion resistant fans, corrosion resistant fans will require special attention.

3.10. Safety Measures

Automatic or remote start is only permitted if this can be done without risk. If safety devices indicate hazardous conditions a remote (re-)start shall not be possible.

In connection with gas tight fans an alarm shall be provided, if increased leakage could represent a hazard.

Emergency stops shall be provided where that may be required.

4. MARKING (Data plate)

Fans shall have data plates supplying the following information:

a) General
- Name and addresses of the manufacturer or responsible vendor
- Mandatory marking
- Designation, type
- Serial number
- Year of construction
- Rating information (mandatory for electrotechnical products: voltage, frequency, power etc., and for machinery intended for use in a potentially explosive atmosphere)

b) Technical data recommended for the safe use of the fan
This can include normal and/or maximum values for the following

- Volume flow rate … in m³/s or m³/h
- Fan pressure … in Pa
- Density … in kg/m³
- Temperature t of the gas handled in °C
- Rotational speed N in r/min
- Absorbed power P in kW
- Acceptable range of impeller blade angles (adjustable pitch impellers)
- Weight in kg

4.1 Signs and warnings
4.1.1 Generally applicable
The direction of rotation and airflow shall be marked on the fan by means of an arrow.

Where necessary suitable warning against operation with hatches or covers removed shall be attached to the fan housing.

4.1.2 Hot gas fans
Suitable warning symbols shall be attached to the fan for hot surfaces.

5. INSTRUCTIONS FOR USE

Details are given for transport and storage, product description, installation, pretesting checks, test runs and commissioning, operation and maintenance.

6. FINAL REMARKS

This paper can not give all rules and recommendations of the standard. Please refer to the integral text of the standard.

© 2000 Hans Witt

S723/003/2000

The revised edition of ISO 5136 – Induct noise measuring standard for air moving devices

W NEISE
Deutsches Zentrum für Luft- und Raumfahrt (DLR), Institut für Antriebstechnik, Berlin, Germany

SUMMARY

Many technical fan installations involve a duct on the inlet side and/or outlet side of a fan. Therefore the need arises for a measurement method for determination of the sound power radiated by fans or other fluid handling sources into a duct. In the paper the problems associated with such a method are discussed. Axial standing waves due to sound reflection from the duct end, acoustic loading of the source, turbulent flow pressures superimposed on the sound field, flow noise generated by flow/microphone probe interaction (self noise), discrimination between sound pressures and turbulent flow pressures, measurement position in the duct in view of higher-order mode sound propagation and directional characteristic of the microphone probe used, and modal distribution of sound power. Early and recent work on the above topics is reviewed. A brief description of the standardized in-duct method ISO 5136:1990 [1] as well as the revised version ISO/DIS 5136:1999 [2] is given.

1. INTRODUCTION

Determination of sound power in a flow duct is of practical interest since many air moving devices are connected to a duct, the most prominent examples being fans. In the late 1960s an international working group was set up (ISO/TC 43/SC 1/WG 3, "Noise from heating, ventilating and air-conditioning equipment") to establish an international measurement standard on "Sound measurement procedure for air moving devices connected to either a discharge duct or an inlet duct". The first standard ISO 5136 [1] was published in 1990. Practical experience indicated that technical problems existed with the standardized measurement procedure and some of the frequency corrections used. In 1996 a new working group ISO/TC 43/SC 1/WG 47 was set up to revise ISO 5136 [1], and in 1999 a Draft International Standard ISO/DIS 5136 [2] was published.

2. DESCRIPTION OF GENERAL PROBLEM

The following general problems exist when one wants to determine the sound power of an air moving device radiated into a duct connected to it:

- The sound power generated by an acoustic source depends on its acoustic load, i.e., the acoustic impedance connected on its inlet and outlet side.
- The sound power propagating from the source down the duct is reflected at duct discontinuities and/or the duct end.
- Above a certain frequency, the sound pressure in the duct is not constant across the duct anymore but depends on the transverse position.
- The microphone in the flow duct is exposed not only to the sound pressures to be measured but also to the unsteady pressures associated with the turbulent flow; as a solution a special windscreen is required.
- In a practical measurement situation, it is difficult to decide whether or not the microphone signal is contaminated by the turbulent flow pressures even when a windscreen is used.

2.1 Axial Standing Waves and Acoustic Loading

Sound reflections occurring at the end of the test duct result in axial standing waves, and as a result the sound pressure to be measured is dependent on the axial location. This problem can be removed by attaching an anechoic termination to the test duct. Many designs have been tested by various authors which are depicted in ISO 5136 [1], [2]. Usage of an anechoic termination provides a well defined acoustic loading which is nearly equal to the characteristic wave impedance ρc (ρ = density, c = speed of sound) and independent of frequency.

Very often it is necessary to use transition ducts to connect the device to be tested to the measurement duct. It is well known that changes in cross section involve sound reflections as well as changes in acoustic loading. In ISO 5136 [1], [2]. the maximum area ratio of transitions is limited to $s_{tr} = S_{large}/S_{small} = 2$ and the minimum length is specified as $L/L_0 \geq s_{tr} - 1$ ($L_0 = 1$ m). This limit is equivalent to the condition $kL \geq 1.65 \cdot (s_{tr} - 1)$ at 90 Hz (k = wave number), see Bolleter *et al* [3].

2.2 Higher-Order Acoustic Mode Sound Propagation in the Test Duct

Only when the sound wavelength is large compared to the cross dimension of a duct is the sound pressure constant over the cross section, and the sound power level L_W transmitted in a duct of cross sectional area S is given by $L_W = L_p + 10 \cdot \lg (S/S_0) + 10 \cdot \lg \{\rho c/(\rho c_0)\}$ where L_p is the sound pressure level; $S_0 = 1$ m^2 and $(\rho c)_0 = 400$ Ns/m^3 are the usual reference values.

At frequencies where higher-order acoustic modes can propagate, the sound pressure amplitude varies across the duct, and determination of the duct sound power requires integration of the sound intensity over the cross section. Sound intensity measurements in the presence of superimposed mean flow are difficult if not impossible due to the sound propagation characteristics in ducts. In modern sound intensity measurement technique, the acoustic velocity is determined via the pressure gradient, more precisely via the pressure difference at two measurement points slightly separated in space. It was shown by Munro & Ingard [4] that "the acoustic intensity cannot be uniquely reconstructed from a measurement of the pressure and pressure gradient at a single point, when there is mean flow", the reason being that there is no unique relationship between pressure gradient and acoustic velocity when the directions of flow and sound propagation are different.

A practical approach for sound power determination in a circular pipe was taken by Barret & Osborne [5]: They specified a radial microphone position in the pipe where the sound pressure is of such magnitude that application of the plane wave equation gives the correct sound power even though higher-order modes are present. Under the assumption that the first ten propagational modes within each given frequency band carry the same acoustic power (mode model of "equal modal sound power", EMSP), this "optimum radial position" is half-way between the pipe axis and the wall ($2r/d = 0.5$), provided a microphone with uniform directivity is used.

Bolleter *et al* [3] found that for a directional sensor like a microphone equipped with a turbulence screen, see the following section, the "optimum radial position" is closer to the pipe wall, where the mode amplitudes are generally larger, to compensate for the microphone directivity. For a slit-tube microphone with 400 mm effective length they recommended the following positions: $2r/d = 0.8$ for pipe diameters from $d = 0.15$ to <0.50 m, and $2r/d = 0.65$ for $d = 0.50$ to 2.0 m. Also, circumferential averaging at the specified radial position was found necessary to obtain a sound pressure amplitude representative for the sound power transmitted in the duct.

2.3 Suppression of Turbulent Flow Pressures
The microphone placed in the duct is subject not only to the acoustic pressures but also to the pressure fluctuations associated with the turbulent duct flow. It was shown by Neise [6] that not the disturbances that are generated by flow/microphone probe interaction are causing the flow noise signal at the microphone but the inherent pressure fluctuations of the turbulent duct flow. To suppress the effect of these turbulent pressures on the microphone, the use of a windscreen is necessary. Friedrich [7] introduced the concept of a long cylindrical windscreen for usage in ducts, and Neise [6] later on suggested a slit-tube design, derived a one-dimensional theory for this design and experimentally verified the theoretical results.

The basic principle of this flow noise suppressor can be explained as follows: A fluctuating pressure field outside the tube excites pressure disturbances along the inner tube wall. From there the pressures are propagated as sound waves to the microphone. If the propagation velocity of the external pressures is different from the speed of sound, the pressures coming from different locations along the tube wall have different phase relationships, and hence the measured pressure is diminished. This effect becomes larger as the ratio of external wavelength to slit-tube length becomes smaller. Thus the microphone will measure the correct amplitude of a plane sound wave travelling in the direction of the tube axis, while the turbulent pressure fluctuations which are convected at a velocity of the order of the flow velocity, $U_c < c$, are sensed to a lesser degree. Therefore the ratio of acoustic pressures to flow noise pressures is increased.

Note that sound waves propagating with or against the mean flow are also sensed to a lesser degree. Further, the axial phase velocity of a sound wave that impinges on the slit-tube microphone at an angle is also different from the speed of sound, and this explains why this probe has a directional characteristic. This feature is important in view of the propagation characteristics of the higher-order acoustic duct modes.

2.4 Discrimination Between Sound Pressures and Turbulent Flow Pressures

In the practical situation of sound measurements in flow ducts one has to ensure that there is a sufficient signal-to-noise ratio between the sound signal and the flow noise signal, even when a windscreen is used. Neise & Stahl [8] presented a procedure to determine the flow noise spectra which involves a comparative measurement using a microphone with nose cone and a microphone with a slit tube. The method is based on the assumption that the sound signal to be measured and the flow noise are mutually uncorrelated and on the experimental observation that there is a difference ΔL_t between the flow noise spectra of a slit-tube microphone and a nose cone microphone.

It was shown by Neise & Stahl [8] that the requirement that the flow noise level at the slit tube microphone be at least $\Delta L_{min} = 6$ dB lower than the sound pressure level reading is equivalent to the condition that the difference between the pressure level readings when using the nose cone and when using the slit-tube must not exceed a maximum value ΔL_{max} which is a function of the flow noise suppression capability ΔL_t of the slit-tube. The steps necessary for this procedure are described in the original paper by Neise & Stahl [8] and also in ISO 5136 [1], [2].

In the latter, another method is described which involves a second sound measurement with a silencer mounted between the source under test and the microphone location. The silencer must have the same cross sectional area and the same length as the replaced part of the test duct. The silencer shall have an insertion loss of at least 10 dB for each frequency band of interest. The requirement for the minimum signal-to-noise ratio of sound to turbulence noise of 6 dB is fulfilled if the average sound pressure level obtained with the silencer in place is at least 5 dB lower than without the silencer.

3. DESCRIPTION OF THE IN-DUCT METHOD ISO 5136 [1]

In ISO 5136 [1], the test duct is of circular cross section and terminated nearly anechoically. The sound power determined under these special conditions is a representative value for actual applications, because the anechoic termination provides an acoustical impedance which lies about midway between the higher and lower impedances encountered in practice and is nearly independent of frequency. The duct terminations are specified in ISO 5136 [1] in terms of the maximum reflection coefficient as a function of frequency.

ISO 5136 [1] is specified for engineering grade accuracy measurements. The range of test duct diameters covered is from 0.15 to 2.0 m, the range of flow velocities 0 to 30 m/s, the maximum swirl angle allowed in the test duct is 15°, and the one-third octave band centre frequency range is from 50 to 10 000 Hz. Each test duct can be used for a limited, specified range of fan sizes by employing conical duct transitions which have to meet certain requirements. Intermediate ducts have to be mounted at the fan inlet and outlet to ensure undisturbed flow conditions. If the non-measured side of the fan is normally ducted in the practical application, a terminating duct with an anechoic termination has to be mounted on this side. If the non-measured side of the fan is normally unducted, no terminating duct is required.

Use of a cylindrical windshield for the microphone (turbulence screen) is prescribed which shall suppress the turbulent pressure fluctuations by at least 10 dB in the frequency range of interest: The directivity characteristic of the microphone with turbulence screen has to be within specified limits. The two procedures described in section 2.4 are incorporated in ISO 5136 [1] to determine whether or not there is a sufficient signal-to-noise ratio of sound to turbulence. The microphone with the windscreen is mounted at a specified radial position such that the measured sound pressure is acceptably well related to the sound power by the plane wave formula, even in the frequency range of higher-order duct modes, compare section 2.2. The radial measurement position r is a function of the test duct diameter d, i.e., $r = 0.8\, d/2$ for $0.15\text{m} \leq d < 0.5$ m and $r = 0.65\, d/2$ for 0.5 m $\leq d \leq 2.0$ m. A circumferential average has to be obtained by measuring at least three evenly spaced azimuthal positions or by a continuous circumferential traverse.

The sound power level in the test duct of cross section S is determined by the circumferentially averaged sound pressure level using the following relation

$$L_W = \overline{L_p} + 10 \cdot \lg (S/S_0) - 10 \lg[\rho c/(\rho c)_0] + C_1 + C_2 + C_3 + C_4 \qquad (1)$$

where C_1 is the free field microphone response correction, C_2 the frequency response correction of the turbulence screen, C_3 the flow velocity correction which accounts for the change in the frequency response of the turbulence screen as a result of the superimposed flow, and C_4 the so-called modal frequency correction which for the fact that the sound pressure measured at the specified radial position does not give exactly the correct sound power in the duct when applying the plane wave formula. C_4 is dependent on the directivity characteristic of the microphone with the turbulence screen because of the propagation angle of the higher-order duct modes.

The head-on frequency response C_2 has to be calibrated under acoustic free-field conditions. Data for C_3 as a function of frequency and flow velocity, and for C_4 as function of frequency and duct diameter are tabulated in ISO 5136 [1]. The C_3-data are based on theoretical investigations by Neise [6], and the data for the modal correction C_4 were computed by Bolleter [9] based on the following assumptions:

- Reflections from the duct end are disregarded,
- non-propagational duct modes can be neglected,
- the total sound power at any one frequency is distributed uniformly over all propagational modes (mode model of equal modal sound power),
- the various duct modes excited by the source are uncorrelated,
- the effect of the mean flow on the sound field in the duct is neglected, i.e., $M = 0$; the only flow effect considered is that on the plane wave sensitivity of the slit-tube (C_3-correction), and
- the directivity characteristic D of the microphone with turbulence screen can be described by an empirical formula.

4. PRACTICAL EXPERIENCE WITH ISO 5136 [1]

In 1984/85 a European round-robin test was carried out to determine the level of uncertainty with which fan sound power levels can be determined following the procedure described in

ISO/DIS 5136.2 [10]. Six laboratories took part in the test (Bolton [11]). Experiments were made with an axial fan and a centrifugal fan. Each laboratory performed the measurements firstly with a reference test rig, which was supplied together with the fans, and secondly with a rig of its own design. The standard deviations for the sound power levels determined for the centrifugal fan for all participating laboratories were found to be within the limits specified in ISO/DIS 5136.2 [10], and similar values were found for the inlet-duct deviations on the axial fan. For the outlet side of the axial fan, the uncertainties were greater than the specified values; this was referred to the amount of swirl in the discharge duct. The results of the round-robin test were used to establish the standard deviation data given in ISO 5136 [1].

Swirl flow often occurs in the outlet ducts of axial fans, in particular when they have no guide vanes, and in these cases increased flow noise levels may arise on the turbulence screen. It was shown by Farzami & Guedel [12], however, that these problems can be overcome by placing a flow straightener between the fan outlet and the measurement plane for the acoustic measurements.

Holste & Neise [13] performed an experimental comparison of the in-duct method ISO 5136 [1] with the free-field method and the reverberation-room method. Six fans with impeller diameters between 450 and 510 mm were used for the experiments. Very good agreement was observed between the results of the free-field method and reverberation-room method.

In the plane wave region, the in-duct method yielded higher levels than the other two methods which is due to the reflection of the sound waves at the fan inlet or outlet, when the duct is removed. In the frequency range of higher-order mode sound propagation, the in-duct sound power levels are lower than the free-space levels. The level difference is frequency dependent with average values of about 3 dB on the inlet side and about 5 dB on the outlet side. Similar results were reported by Bolton [14]. In this frequency regime, where the wavelength is small compared to the cross dimensions of the duct, effects of sound reflection and acoustic loading are unlikely to play a significant role, and one would expect all test methods to deliver the same result. Since the free field tests and the reverberation room tests yielded consistent results, it appeared that the reason for this discrepancy lies in the in-duct method. As a possible cause, Holste & Neise [13] named the two frequency correction terms C_3 and C_4 in ISO 5136 [1]. The data for the flow velocity correction C_3 are the result of theoretical investigations by Neise [6] which have been verified experimentally in the meantime (Neise [15]).

As was mentioned above, the modal correction C_4 accounts for the directivity of the microphone with the turbulence screen with respect to the propagation angle of the higher-order duct modes. The assumptions made for calculating the C_4-data given in ISO 5136 [1] were described in chapter 3.

Michalke [16] showed that higher values for the modal correction C_4 are in fact obtained when the theoretical slit-tube model described by Neise [6] was used to calculate the microphone probe directivity, rather than the empirical relation used by Bolleter [9]. Theoretical studies further revealed that the modal correction is a function of the mean flow velocity (Michalke [16]), as well as of the flow direction relative to that of sound propagation. For that reason, an experimental and theoretical study was started at DLR in Berlin to obtain more accurate data for the modal correction to be implemented in the standard. This study was supported by the Deutsche Forschungsgemeinschaft (DFG, German Research Foundation).

5. DETERMINATION OF MODAL CORRECTIONS

5.1 Experimental Determination of Modal Corrections

To determine modal correction data for a specified measurement path in a circular duct, one needs to know or to be able and measure the actual ("true") sound power transmitted in the duct. It was shown by Munro & Ingard [4] that the modern sound intensity technique, where the acoustic velocity is determined via the pressure difference at two slightly displaced points, cannot be applied in the higher-order mode frequency regime in the presence of flow because there is no unique relationship between pressure gradient and axial acoustic velocity when directions of flow and sound propagation are different.

Many investigators have tried to find other ways to determine the sound power in a duct. A thorough review of these papers was given by Arnold [17]. Here only the most recent ones are outlined: Michalke [18], [19] proposed a method for sound power determination which involves the area-averaged cross spectra of the sound pressure at various locations in the circular duct. This method does not require any assumptions with regard to the intermodal coherence of duct modes, however, it yields reliable data only for the non-dimensional frequency range up to $kR = 5.3$. Arnold [17] refined Michalke's approach to extend the frequency range to about $kR = 30$ and was able to resolve up to 100 acoustic modes. His results also enabled him to decide that the mode model of "equal modal energy density" (EME) describes the modal distribution of the sound energy generated by ventilating fans best.

None of the experimental methods studied in the past can be used to determine the "true" sound power in the entire frequency range of practical interest, i.e., the standardized in-duct method covers a non-dimensional frequency range up to $kR = 183$ (10 kHz in a 2 m diameter pipe). Hence, one needs to find other ways to determine modal correction data, and in the following a theoretical approach is described.

5.2 Theoretical Determination of Modal Corrections

Arnold *et al* [20] re-calculated the modal correction data based on Michalke's [18], [19] theory for a microphone equipped with a "standard" turbulence screen (slit-tube) placed at the radial position specified in ISO 5136 [1] under the following assumptions:

- Reflections from the duct end are disregarded;
- non-propagational duct modes can be neglected;
- different mode models are possible, e.g., equal modal sound power, EMSP, and equal modal energy density, EME (EME is used for the data given in [2], because it represents the sound field radiated by a fan better than EMSP, see [17]);
- correlated or uncorrelated higher-order duct modes can be assumed (in [2] uncorrelated modes are chosen, because they approximate an average over all degrees of intermodal correlation, see [17]);
- the effect of superimposed mean flow on sound propagation is taken into account; and
- the directivity characteristic of the turbulence screen is described by the theoretical slit-tube model proposed by Neise [6].

The results of this calculation showed that the modal correction C_4 is in fact a function of the superimposed flow velocity, see Figure 1. On the outlet side (M > 0), the new modal correction data are somewhat larger than the data in ISO 5136 [1], and on the inlet side (M < 0) the modal correction can assume negative values, depending on frequency and flow velocity.

Arnold *et al* [20],[17] proposed that instead of using the two frequency corrections C_3 and C_4, which are both functions of the mean flow velocity, a combined flow velocity and modal correction C_{34} should be introduced which accounts for the effect of mean flow on the acoustic response function of the turbulence screen as well as on the propagation of higher-order sound waves in the test duct.

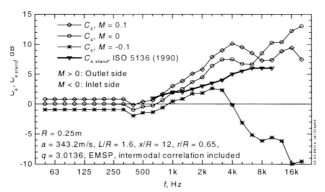

Figure 1: Modal correction C_4 as a function of flow velocity (after Arnold *et al* (1995)).

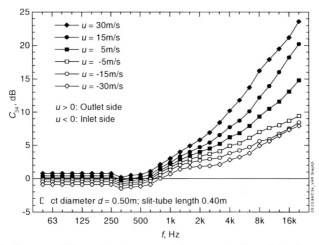

Figure 2: Combined flow velocity and modal correction for a 0.4 m long slit-tube in a 0.5 m diameter circular duct (after Arnold [17]).

In Figure 2 computed results are shown for the combined correction for a standard length slit-tube in a pipe of 0.5 m diameter. The frequency dependence of the C_{34}-data is much smoother than that of the former C_3- and C_4-data.

Arnold [17] compared the sound power spectra determined by using the in-duct method with the new C_{34}-correction with the "true" sound power determined by using his improved correlation method (see section 5.1) and found excellent agreement. He also showed that with the new C_{34}-correction a much better agreement between the in-duct sound power levels and the free-field sound power levels is obtained.

6. REVISED IN-DUCT METHOD ISO/DIS 5136 [2]

A schematic layout of a test setup for the revised in-duct method is depicted in Figure 3. The test duct diameter range is from 0.15 m to 2.0 m. With the transition pieces allowed, the range of fan inlet equivalent diameters which can be tested is from 0.104 m to 2.0 m, the range of fan outlet equivalent diameters is from 0.104 m to 2.390 m. Test procedures for smaller and larger fans are described in two appendices. All ducts connected to the test fan have to be terminated anechoically.

The range of flow velocities in the test duct is now from 0 to 40 m/s. Three different types of windshield are allowed: Foam ball (up to 15 m/s), nose cone (up to 20 m/s), and turbulence screen (up to 40 m/s). The latter is the preferred probe, and the uncertainties specified in the standard apply only when the turbulence screen is used. No uncertainty information is given for the cases when foam ball or nose cone are used. The radial measurement position of the microphone with turbulence screen is the same as in ISO 5136 [1].

Compared to the previous version of the method, the ducting arrangements on the inlet and outlet side were modified to enable simultaneous aerodynamic performance testing according to ISO 5801 [2]. A "star type" flow straightener is mounted upstream of the outlet test duct. The straightener is necessary for aerodynamic fan performance testing, and it eliminates the negative effect of swirl flow on the microphone windshield. The outlet duct noise shall be measured with and without the flow straightener in position. Of the sound pressure level readings taken, the lowest shall be considered to represent the true sound pressure in the test duct, for each one-third octave band of interest. The in-duct sound power level is now governed by the relation

$$L_W = \overline{L_p} + 10 \cdot \lg (S/S_0) - 10 \cdot \lg[\rho c/(\rho c)_0] + C_1 + C_2 + C_{34} \qquad (2)$$

where C_{34} is the new combined flow velocity and modal correction. Data for C_{34} are given in ISO/DIS 5136 [2] as function of frequency for the test duct diameter and flow velocity range covered; for information, C_{34}-data are also given for flow velocities up to 60 m/s and for the extended duct diameter range described the two appendices mentioned above.

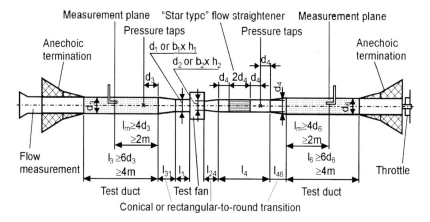

Figure 3: Test setup for the revised in-duct method ISO/DIS 5136 [2].

7. CONCLUSIONS

The general problems involved in sound power determination in circular flow ducts have been discussed. Results from published literature relevant to the development of a standardized measurement procedure have been reviewed and the standardized in-duct method ISO 5136 [1] has been outlined.

With the revision of the in-duct method ISO/DIS 5136 [2], technical problems have been removed that became evident in previous experiments. The flow velocity correction C_3 and the modal correction C_4 used in ISO 5136 [1] were replaced by a new combined flow velocity and modal correction C_{34}. Recent tests with the new correction proved that very good agreement is now obtained in the frequency region of higher-order sound propagation between in-duct sound power levels and free-field sound power levels.

The test setups were modified to enable simultaneous acoustic and aerodynamic performance testing. A star-type flow straightener is used upstream of the discharge test duct to remove swirling flow. The in-duct test procedure requires that the sound pressure be measured in an anechoically terminated test duct. Due to this, the method is largely insensitive to environmental conditions and background noise problems. Only three microphone positions in the duct are required or one circumferential microphone traverse. The method has been tested specifically for fans in a European round-robin test (Bolton [11]) and was found very practical and to give reproducible results. The in-duct method is of "engineering grade measurement accuracy" when the turbulence screen (slit-tube) is used as a microphone wind shield.

8. REFERENCES

[1]   ISO 5136, Acoustics - Determination of sound power radiated into a duct by fans - In-duct method, (1990).
[2]   ISO/DIS 5136, Determination of sound power radiated into a duct by fans and other air

moving devices - In-duct method (Revision of ISO 5136:1990), (1999).

[3] Bolleter, U., Cohen, R. & Wang, J., Design considerations for an in-duct sound power measuring system. *Journal of Sound and Vibration* **28** (1973), 669–685.

[4] Munro, D.H. & Ingard, K. U., On acoustic intensity measurements in the presence of mean flow. *Journal of the Acoustical Society of America* **65** (1979), 1402–1406.

[5] Barret, A.J. & Osborne, W.C., Noise measurement in cylindrical fan ducts. *Journal of the Institution of Heating and Ventilating Engineers* **28** (1960)**,** 306–318.

[6] Neise, W., Theoretical and experimental investigations of microphone probes for sound measurement in turbulent flow. *Journal of Sound and Vibration* **39** (1975), 371–400.

[7] Friedrich, J., Ein quasischallunempfindliches Mikrofon für Geräuschmessungen in turbulenten Luftströmungen. *Technische Mitteilungen Rundfunk- und Fernsehtechnisches Zentralamt* **11** (1967), 30–33.

[8] Neise, W. & Stahl, B., The flow noise level at microphones in flow ducts. *Journal of Sound and Vibration* **63** (1979), 561–579.

[9] Bolleter, U., The measurement of sound power in circular ducts with a directional microphone in the presence of cross modes. ISO/TC 43/SC 1/WG 3 N62 (1980).

[10] ISO/DIS 5136.2, Determination of sound power radiated into a duct by fans and other air moving devices - In-duct method *Draft International Standard* (1985)

[11] Bolton, A. N., Intercomparison of in-duct fan noise measurements. Report EUR 13890 EN, Comm. Europ. Communities, Community Bureau of Reference, Brussels, Belgium (1992).

[12] Farzami, R. & Guedel, A., Influence of a straightener on in-duct sound power measurements. *Proc. Int. Conf. Fan Noise*, Senlis, France, CETIM (1992), 375–380.

[13] Holste, F. & Neise, W., Experimental comparison of standardized sound power measurement procedures for fans. *Journal of Sound and Vibration* **152** (1992), 1–26.

[14] Bolton, A. N., Fan noise installation effects. *Proc. Int. Conf. Fan Noise*, Senlis, France, CETIM, (1992), 77–88.

[15] Neise, W., Experimental determination of the acoustic sensitivity of a microphone turbulence screen under mean flow conditions. *Proc. Inter-noise '91*, Sydney, Australia (1991), 1013–1016.

[16] Michalke A., On the sensitivity of a slit-tube probe to the higher-order acoustic modes and its consequences. *Proc. DGLR/AIAA 14*[th] *Aeroacoustics Conference 1992*, Aachen, Germany. DGLR-Bericht **92-03** (1992), 163–168.

[17] Arnold, F., Experimentelle und numerische Untersuchung zur Schallleistungsbestimmung in Strömungskanälen. *VDI-Fortschritt-Berichte Reihe 7* Nr. **353,** VDI-Verlag GmbH Düsseldorf (1999).

[18] Michalke, A., On the propagation of sound generated in a pipe of circular cross-section with uniform mean flow. *Journal of Sound and Vibration* **134** (1989), 203–234.

[19] Michalke, A., On experimental sound power determination in a circular pipe with uniform mean flow. *Journal of Sound and Vibration* **142** (1990), 311–341.

[20] Arnold, F., Hahn, J., Michalke, A. & Neise, W., Zur Schallleistungsbestimmung in Strömungskanälen mit Schlitzrohr-Sonden. *In Fortschritte der Akustik, DAGA'95* (1995), 531–534.

[21] ISO 5801, Industrial fans - Performance testing using standardized airways (1997).

S723/005/2000

The new balancing and vibration standards for fans

W T W CORY
Woods Air Movement Limited, Colchester, UK

1.   INTRODUCTION

The early fans ran at very low peripheral speeds. In consequence the out of balance forces were small and the need for the balancing of rotating parts, rarely if ever became a necessity. Even then, it was common to statically balance the impeller and shaft on knife edges and this sufficed for many decades.

As pressure requirements became greater, so also did the peripheral speeds, such that the necessity for more sophisticated forms of balancing became essential. Where the volumetric flowrates as well as the pressures were high, the width to diameter ratio of impellers became greater with the consequent probability of couple effects being set up.

It is no surprise therefore, that fan manufacturers were among the first users of dynamic balancing machines and that the craft of balancing became a highly valued ability of their skilled assembly fitters. Since that time, many other sources of fan vibration have been recognised such that the need for recognised industry standards has become obvious.

Rather more recently, electric motors, steam turbines, compressors and many other forms of rotodynamic machinery have been invented. These with their higher rotational speeds have had to be balanced to a fine degree. In consequence their manufacturers have been influential in the formulation of standards based on the work of Rathbone and others. These give numerical values to the out-of-balance forces and assign grades for particular types of machine usually based on kW rating.

It is a matter of regret that the fan industry was not represented on the drafting committees in any strength. The appropriate grades are not therefore always acceptable for fans when compared with other rotating machinery of equivalent power because:

a)   the static parts are much lighter
b)   the heights from support level to rotating shaft centreline are often much higher
c)   the ratio of running speed to first critical speed is very different

d)   the ratio of rotating parts weight to static parts weight can be much higher

e)   fans often have to handle substantial quantities of solid or liquid matter with consequent erosive or corrosive effects.

Whilst, therefore, the mathematics, units and method of grading incorporated in generic standards such as ISO1940, ISO2372 (now withdrawn) and ISO10816 were, or can be, accepted with enthusiasm, it was felt that machine specific guidelines were a necessity. It is under these constraints that ISO14694 and ISO14695 have been written.

2.   THE NEED FOR STANDARDS

ISO14694 and ISO14695 address the need of both users and manufacturers of fans for technically accurate but uncomplicated information on balance precision and vibration levels.

Vibration is an important parameter in the description of the performance of a fan. It gives an indication of how well it has been designed and constructed. Possible operational problems associated with inadequate support structures, machine deterioration etc can be foreseen and obviated.

Vibration measurements may therefore be required for a variety of reasons of which the following are the most important examples:

i)   design/development evaluations
ii)   in situ acceptance testing
iii)   as information for a condition-monitoring or machinery health programme
iv)   to inform the designer of associated supporting structures, foundations, ducting systems etc., of the residual vibration, which will be transmitted by the fan into these.
v)   As a quality assessment at the final inspection stage

A considerable amount of information can be generated from tests carried out in accordance with the standards. Not of all of this is either necessary or appropriate when carried out for one of the specific reasons detailed above. For example, quality grading may only require a minimum of information to be specified. It is especially important to remember that vibration testing can be extremely expensive, often greater than the fan itself. If specified it will inevitably increase the price. Only where the functioning of the installation could be affected should discrete frequency or band limitations be imposed. The number of test points should be minimised to the duty range. Readings at the fan bearings are usually sufficient for normal quality gradings especially as they relate to impeller balance.

It is important to remember that factory tests are usually conducted with the fan unconnected to a ducting system such that its aerodynamic duty will differ considerably from that during normal operation. The temporary supports and foundations will inevitably be of different mass and stiffness to those on site. Accordingly the readings are of use only with vibration 'filter-in' when the residual unbalance can be measured. Overall vibration severity (filter-out') can only be obtained in situ with the proper foundations, with the proper ductwork fitted and with the proper operating conditions.

3.   SOURCES OF VIBRATION

It will be seen that it is impossible to completely eliminate all vibration as this arises from the dynamic effects of residual out-of-balance, misalignment, eccentricity, clearances, rubbing or rolling contacts, the additive effects of essential manufacturing tolerances, looseness between mating parts, aerodynamic fluctuating forces etc. Often the vibrations from these sources may be small, but can excite the resonant frequencies of the stationary parts such as casings or bearing pedestals. Where the fan is directly driven by an electric motor, electromagnetic disturbances may be present leading to the possibility of further resonant conditions and consequent vibration.

4.   DEFINITIONS OF VIBRATION

Vibration may be defined as the periodic motion in alternately opposite directions about a reference equilibrium position. The number of complete motion cycles which take place during unit time is called the frequency. In Système International units this frequency is measured in cycles/second or Hertz (Hz).

The motion can consist of a single frequency, as with a tuning fork. With a fan however there are likely to be many motions taking place simultaneously at different frequencies. These various motions can be identified by frequency analysis - or the plotting of a graph showing vibration level against frequency.

5.   VIBRATION MEASURING PARAMETERS

There are three properties of a vibrating element which can be measured (Fig. 1) in addition to the frequency at which the vibration takes place. Each may be of value to particular individuals be they designers, fitters, installers or users:

a)   **Displacement**, or the size of the movement is of importance where running clearances have to be maintained for efficient performance, or where contact between stationary and rotating surfaces could take place. Most weight is given to the low frequency components of a vibration. Only below about 10Hz is it usual to use displacement as the key to the cause.

b)   **Velocity,** which is directly proportional to a given energy level and therefore where low and high frequencies are equally weighted. The disturbing effects on people and other equipment are by experience related to velocity. For this reason vibration severity is usually specified in most standards as a velocity in mm/s and for the user and consultant is the preferred parameter.

c) **Acceleration,** which is a measure of the forces and stresses set up within a fan and/or motor, or between these and the foundation or ductwork. Weighted towards the higher frequencies and should be used where such components exist. Of most interest to fan and system designers, but will also be of value in measuring the degree of fatigue in bearings.

6. SIMPLE HARMONIC MOTION

The three parameters described about are mathematically connected in the case of a simple harmonic or sinusoidal vibration such as that produced by out-of-balance. The displacement 'e' is proportional to $\sin \omega t$ where ωt is an angle which goes through $360°$ in one vibratory cycle. Angular velocity (or circular frequency) ω is equal to $2\pi f$ where f is the frequency in Hertz, or $\dfrac{2\pi N}{60}$ for balance problems where N equals r/min.

The other properties are also sine waves, the velocity 'v' having a $90°$ phase lead (one quarter or a cycle with respect to time) whilst acceleration 'a' is advanced by half a cycle i.e. a $180°$ phase lead. This is shown in the equations below:

$$\text{Displacement} \quad e \quad = e_{peak} \sin \omega t$$

$$\text{Velocity} \quad v \quad = \omega e_{peak} \sin\left(\omega t + \frac{\pi}{2} \right)$$

$$\text{Acceleration} \quad a \quad = \omega^2 e_{peak} \sin(\omega t + \pi)$$

These three parameters vary during the cycle. Table 1 gives their values with respect to e_{peak}.

No.	Radians	Degrees	Displacement	Velocity	Acceleration
1	0	0	0	ωe_{peak}	0
2	0.25π	45	$1.71 \times e_{peak}$	$0.71 \times \omega e_{peak}$	$-0.71 \times \omega^2 e_{peak}$
3	0.5π	90	e_{peak}	0	$-\omega^2 e_{peak}$
4	0.75π	135	$0.71 \times e_{peak}$	$-0.71 \times \omega e_{peak}$	$-0.71 \times \omega^2 e_{peak}$
5	π	180	0	$-\omega e_{peak}$	0
6	1.25π	225	$-0.71 \times e_{peak}$	$-0.71 \times \omega e_{peak}$	$\omega^2 e_{peak}$
7	1.5π	270	$-e_{peak}$	0	$-0.71 \times \omega^2 e_{peak}$
8	1.75π	315	$-0.71 \times e_{peak}$	$0.71 \times \omega e_{peak}$	$-0.71 \times \omega^2 e_{peak}$
9	2π	360	0	ωe_{peak}	0

Point in cycle (Radians, Degrees under it for columns No., Radians, Degrees)

Table 1. Values of parameters expressed as function of peak displacement

7. WHICH VIBRATION LEVEL TO MEASURE

It will be seen that all these quantities vary with time. For analytical purposes it is desirable to reduce them to single figures and those for displacement are shown in (Fig. 2).

The **peak-to-peak** value indicates the total excursion of the wave and is useful in calculating maximum stress values or determining mechanical clearances.

The **root-mean-square** value is probably the most important measure because it takes account of the cycle time and gives an amplitude value which is directly related to the energy content and therefore the destructive capabilities of the vibration. For sine wave vibrations e.g. out of balance

$$e_{rms} \times \sqrt{2} = e_{peak}$$

Peak and average values may also be calculated but have a limited value.

$$e \sqrt{\frac{1}{T} \int_o^T e^2(t)\, dt}$$

$$e_{av} = \frac{1}{T} \int_O^T |e|\, dt$$

Velocities and accelerations are given in similar terms and the root-mean-square velocity is especially important as it is used in ISO 10816 as the measure of vibration severity in the range 600 to 12000 r/min.

Again for a sinusoidal vibration:-

$$v_{rms} \times \sqrt{2} = v_{peak}$$

It must be emphasized that the relationships connecting root-mean-square and peak values only apply to sine-waves. Vibrations arising from certain other sources e.g. rough rolling element bearings or air turbulence may not follow this form. Consequently the equivalents in Table 1 will not hold and the acceleration values especially may be much higher.

Where sine-wave conditions do exist, by taking time-average measurements the effects of phase may be ignored and:-

Displacement $\quad e = \dfrac{a}{4\pi^2 f^2} = \dfrac{v}{2\pi f} = \int v\, dt$

Velocity $\quad v = \dfrac{a}{2\pi f} = \int a\, dt$

Acceleration $\quad a = 2\pi f v$

The values of e, v or a may be either root-mean-square or peak as applicable.

Whilst all ISO standards use the r.m.s. value, it should be noted that much North American vibration measuring equipment is calibrated for peak levels.

8.   UNITS OF MEASUREMENT

All these parameters should be measured in Système Internationale customary units, and these are adopted in ISO14694 and ISO14695.

In North America however inch-pound units are still often used and these also are shown in Table 2.

Property	S.I. = Metric	Imperial = U.S.
Displacement	μm = 0.001mm	thous = mils = 0.001 in
Velocity	mm/s	In/s
Acceleration	m/s^2	g's (1g = 32.17ft/s^2)
Frequency	Hz = cyc/sec	cyc/min

Table 2. Units used in vibration measurement

9.   DECIBELS AND LOGARITHMIC SCALES

Frequency is almost invariably plotted logarithmically to keep the scale length down to a reasonable size. It also results in the lower frequency part being expanded whilst the high frequency part is compressed. A constant percentage resolution is obtained over the whole chart.

In like manner logarithmic scales may be used for plotting vibration velocities and accelerations. As the absolute values can vary enormously, and to enable vibration levels to be easily compared, decibel scales are often used. From our knowledge of noise levels it is appreaciated that the decibel (dB) is the ratio of one level with respect to a reference level. It therefore has no dimensions. To obtain absolute values the reference level must be known. It is an unfortunate fact that there are two commonly used sets of reference levels - those recommended in ISO 1683 and those used in defence products. These are set out in Table 3.

Property	Definition	ISO	Defence
Acceleration	$L_a = 20 \log \left[\dfrac{a}{a_O} \right]^{A \, dB}$	$A_O = 10^{-6} \, \text{m/s}^2$	$A_O = 10^{-5} \, \text{m/s}^2$
Velocity	$L_v = 20 \log \left[\dfrac{v}{v_O} \right]^{V \, dB}$	$V_O = 10^{-9} \, \text{m/s}$	$v_O = 10^{-8} \, \text{m/s}$

Table 3. Vibration definitions for decibel scales

For the same absolute values, the ISO levels will therefore be 20dB higher than those used in the defence industry. It is important to remember this when comparing figures and to make sure that the appropriate reference level is given.

10. FAN RESPONSE

The fan and its parts may be likened to a spring-mass system and an understanding of this fact is useful in resolving many vibrational problems. It is also of importance in revealing the causes of resonance.

Every fan will have three basic properties:-

a) Mass 'm' measured in kg. The force due to the mass of the system is an inertia force or a measure of the tendency of the body to remain at rest.

b) Damping 'C' is the damping force per unit velocity of a system. It is a measure of the slowing down of vibrations and is given in Newton sec/mm.

c) Stiffness 'k' is a measure of the force required to deflect part of the fan through unit distance. Measured in Newton/mm.

The combined effects of these restraining forces determine how a fan will respond to a given vibratory force e.g. unbalance. Thus we may state that:

$$m\,\ddot{e}_p + C\,\dot{e}_p + k\,e_p \qquad\qquad = M_u\,\omega^2 r \sin(\omega t - \varnothing)$$

$$= M\omega^2 e\, e \sin(\omega t - \varnothing)$$

or $\quad - m\,e_p\omega^2 \sin\omega t + C\,e_p\omega \sin(\omega t + \pi/2) + k\,e_p = M_u\omega^2 r \sin(\omega t - \varnothing)$

$$= M\omega^2 e \sin(\omega t - \varnothing)$$

where

e	=	displacement of centre of gravity from centre of rotation	m
e_p	=	displacement of part due to vibratory force	m
M	=	mass of rotating parts	kg
M_u	=	mass of residual unbalance	kg
r	=	distance of unbalance from rotating centre	m
$\varnothing$	=	phase angle between exciting force and actual vibration	rad
ω	=	angular velocity	rad/s
m	=	total mass or mass being considered	kg

or

Inertia force + Damping force + Stiffness force =, Vibratory force

It will be seen that the three restraining forces are not working together, indeed the inertia and stiffness forces are 180° out of phase. There will be a certain frequency where they are equal and will cancel each other out. Then there will only be the damping force (which is 90° out of phase) to oppose the vibratory force.

Consider a system with no damping, then

$$-m\,e_p\omega^2\sin\omega t + k\,e_p\sin\omega t = M\omega^2 e\,\sin(\omega t - \varnothing)$$

The displacement of the system due to the vibratory force can be expressed as

$$e_p = \frac{M\omega^2 e\sin(\omega t - \varnothing)}{(k - m\omega^2)\sin\omega t}$$

The displacement, e_p will become infinitely large for any value of t, when the expression $k - m\omega^2$ becomes zero, thus the critical rotational frequency ω_c at which this occurs is

$$\omega_c = \sqrt{(k/m)}$$

or more specifically for fans, the critical speed N_c rev/min is

$$N_c = \omega_c \times 60/(2\pi) = 60/(2\pi) \times \sqrt{(k/m)}$$

This condition, known as resonance, can cause high levels of vibration, and although minimised by the optimisation of damping, should be avoided by operating at a speed well away from the critical speed.

All fans together with their supporting bases consist of a number of different spring mass systems, each having its own natural frequency possible with various degrees of freedom and a different resonant frequency for each. Whilst unbalance is usually the major exciting force, there will be numerous other sources such that resonance can be a common problem. In these other cases, the force due to out of balance would be replaced in the equation by those due to electromagnetic, aerodynamic or other forces as appropriate.

11.   BALANCING

Balancing is the process of improving the distribution of mass in an impeller so that it can rotate in its bearings without producing unbalanced centrifugal forces.   Perfection is impossible and even after balancing there will be residual unbalance, its magnitude being dependent on the machinery available and the quality necessary for the application.

Until recently the only relevant standard for the balancing of rotating elements was taken to be ISO1940 in its various editions.   These gave recommendations for various rotor groups to avoid gross deficiencies or unattainable requirements.   If the recommended limits were followed, satisfactory running could be expected.   The grade for fan impellers given in

ISO1940 is G6.3 but this is seen in ISO 14694 as being too general and recommended limits are given according to the application and the driver power (Table 4 and 5).

Application	Examples	Driver Power kW Limits		Fan Application Category BV
Residential	Ceiling fans, attic fans, window AC		.15	BV-1
		>	.15	BV-2
HVAC & Agricultural	Building ventilation and air conditioning; commercial systems		3.7	BV-2
		>	3.7	BV-3
Industrial Process & Power Generation etc.	Baghouse, scrubber, mine, conveying, boilers, combustion air, pollution control, wind tunnels		300	BV-3
		>	300	See ISO 10816-3
Transportation & Marine	Locomotive, trucks, automobiles		15	BV-3
		>	15	BV-4
Transit/Tunnel	Subway emergency ventilation, tunnel fans, garage ventilation		75	BV-3
		>	75	BV-4
	Tunnel Jet Fans		ANY	BV-4
Petrochemical Process	Hazardous gases, process fans.		37	BV-3
		>	37	BV-4
Computer Chip Manufacture	Clean rooms	ANY		BV-5

Table 4.   An Application Categories

Note: *This standard is limited to fans below approximately 300kW. For fans above this power refer to ISO 10816-3. However, commercially available standard electric motor may be rated at up to 355 kW (following an R20 series. Such fans will be accepted in accordance with this standard.*

Note: *This table does not apply to the large diameter (typically 2800mm to 125mm diameter) lightweight low speed axial flow fans used in air cooled heat exchanges, cooling towers, etc. The balance quality requirements for these fans shall be G16 and the fan application category shall be BV-3.*

Fan Application Category	Balance Quality Grade for Rigid Rotors/Impeller
BV –1	G 16
BV –2	G 16
BV -3	G 6.3
BV -4	G 2.5
BV -5	G 1.0

Table 5.   Balance quality grades

Note: *In fan application category BV-1 there may be some extremely small fan rotors weighing less than 224 grams. In such cases, residual unbalance may be difficult to determine accurately. The fabrication process must ensure reasonably equal weight distribution.*

An unbalanced impeller will create forces at its bearings and foundations and the complete fan will vibrate. At any given speed the effects depend on the proportions and mass distribution of the impeller as well as the stiffness of the bearing supports. In the past residual unbalance has been resisted by massive supports. Now, it is recognised that a preferable solution is to reduce this unbalance so that unnecessary weight need not be added to the bearing pedestal.

For narrow impellers (width less then 20% of diameter) the static unbalance is of primary importance. Two unbalances (in different planes) in the same direction usually cause a greater disturbance than two equal unbalances in opposite directions. With wider impellers (width up to 50% of diameter) such effects become of importance.

Static unbalance, sometimes called force or kinetic unbalance, can be detected by placing the impeller on parallel knife-edges. The heavy side will swing to the bottom. Correction weight can be added or removed as required and the part is considered statically balanced when it does not rotate on knife-edges regardless of the position in which it is placed (see Fig. 3).

Dynamic unbalance is a condition created by a heavy spot at each end of the impeller but on opposite sides of the centreline. Unlike static unbalance, dynamic unbalance cannot be detected by placing on knife-edges. It becomes apparent when the impeller is rotated and can only be corrected by making balance corrections in two planes (see Fig. 4).

An impeller which is dynamically balanced is also in static balance. Thus there is no need for the two operations where a dynamic balancer is used, despite the many specifications calling for both.

In general, the greater the impeller mass, the greater the permissible unbalance. It is therefore possible to relate the residual unbalance U to the impeller mass m. The specific unbalance $e = \dfrac{U}{m}$ is equivalent to the displacement of the centre of gravity where this coincides with the plane of the static unbalance. Practical experience shows that e varies inversely with the speed N over the range 100 to 30000 rev/min for a given balance quality. It has also been found experimentally that eN = constant (see Fig. 5). The Quality grades which have been establish permit a classification of requirements.

11.   VIBRATION PICKUPS

From the information given so far it will be appreciated that exactly how the vibration is measured and the equipment used becomes of prime importance. The actual 'pickup' or transducer is a sensing device which converts the mechanical vibration into electrical energy. Many different types exist but those most commonly used are as follows:-

a)   **Seismic velocity pickup:** this consists of a coil of wire supported by springs in a magnetic field created by a permanent magnet which is part of the case. For details of the construction see Fig. 6. When it is held against or attached to a vibrating machine, the permanent magnet, being attached to the case, follow the vibratory motion. The coil of wire or conductor being supported by springs of low stiffness remains stationary in space. Thus the conductor is moving through a magnetic field and a voltage is therefore induced. The voltage generated is directly proportional to the velocity.

b) **Piezoelectric accelerometer:** consists of a mass rigidly attached to certain crystal or ceramic elements which when compressed or extended produce an electrical charge (see Fig. 7). The voltage generated by the element is proportional to the force applied and since the mass of the accelerometer is a constant, is proportional to the acceleration. As acceleration is a function of frequency squared they are most sensitive to high frequency vibration.

Requirements for sensitivity and deviation of transducers are given in ISO 14695.

12. VIBRATION ANALYSERS

It is not the intention of this paper to be a technical manual of vibration measuring equipment. Suffice it to say that just as the voltages generated are a function of the property being measured, so the analyser to which they are attached by cable, can reconvert the signals back to velocity or acceleration. Furthermore, due to the mathematical relationships which exist, the addition of an integrator in the circuitry allows the other vibration properties to be obtained. Low and high pass filters are included, and these can be adjusted to limit the frequency range to that of interest or examination, whilst linear to logarithmic converters enable the signal to be displayed correctly. Output sockets are also provided so that a complete vibration signature over the full frequency spectrum can be obtained and displayed on a chart recorder, oscilloscope or tape recorder.

13. VIBRATION ENERGY

The force causing the vibration of a fan has already been established as:

$$Vibratory\ force = M\omega^2 e \sin(\omega t - \phi)$$

Since vibration is assessed in linear terms along an axis, it is useful to express the vibratory force as a linear vector. The expression in terms of an rms value becomes:

$$Linear\ vibratory\ force = 1/\sqrt{2}\ M\omega^2 e$$

Similarly the rms displacement of the rotating masses is given as $1/\sqrt{2}\ M\omega^2 e$. Thus the kinetic energy of the rotating masses, or the potential to induce vibratory motion, is given as the product of the applied force and the displacement due to the force i.e.:

$$Vibration\ energy = {}^1/_2\ M(\omega e)^2$$

From the derivation of G grades in clause 8.2, the expression of vibration energy can be rewritten:

$$Vibration\ energy = {}^1/_2\ MG^2$$

14.   THE CONTENTS OF ISO 14695 AND ISO 14694

Two standards have been prepared for the measurement and evaluation of fan vibration. It is emphasised that these are interdependent and each should be read in conjunction with the other. Unfortunately, the numbering of the standards is perhaps illogical, as the measurement should, it is suggested, precede the specification.

They are titled as follows:

ISO 14695   Industrial fans - Methods of measurement of fan vibration
ISO 14694   Industrial fans - Specification for balance quality and fan vibration

The measurement standard is largely based on BS 848 Part 6 : 1986 which in turn was based on a methods used by a number of Navies. For a direct driven fan it recommends suspension by rubber ropes. The number and size of these ropes is determined such that they are extended by 50% beyond their original length when subjected to the weight of the fan. This ensures a natural frequency of the suspension system of below 5Hz. Thus the measured fan vibration when running will be unaffected by external sources (Fig. 8).

However examples are also given for other types of resiliently mounted fans so that readings of the fan vibration may be recorded at the same time as a performance test in accordance with ISO 5801. How the transducers shall be attached and to what positions is specified in some detail according to the usage of the information obtained (Figs. 9, 10 and 11).

Based on AMCA 204 and the earlier draft BS 848 Part 7, ISO 14694 is concerned with recommended balance qualities and vibration limits for all types and conditions of industrial fans. It is recognised that the design and structure of a fan and its intended application are important criteria in determining applicable and meaningful balance grades and vibration levels. In section 11 of this paper it was noted that categories had been established for balancing. These same BV categories have also been assigned vibration limits measured in the preferred velocity units mm/s. To take account of North American usage they are given as both peak and rms values. As noted previously, those conducted at the manufacturers works, are with 'filter-in' i.e., at rotational frequency (Table 6).' Those measured in-situ are for 'filter-out' (Table 7) and will depend on whether the fan is rigidly or flexibly mounted. They equate to the vibration severity as given in ISO 10816.

Fan Application Category	Rigidly Mounted mm/s		Flexible Mounted mm/s	
	peak	rms.	peak	rms.
BV-1	12,7	9,0	15,2	11,2
BV-2	5,1	3,5	7,6	5,6
BV-3	3,8	2,8	5,1	3,5
BV-4	2,5	1,8	3,8	2,8
BV-5	2,0	1,4	2,5	1,8

Note 1: *Refer to Annex A for conversion of velocity units to displacement or acceleration units for filter-in readings.*

Note 2: *The rms. Values given in table 4 are preferred. They are to a rounded R20 series as specified in ISO 10816-1. Peak values are widely used in North America. Being made up of a number of sinusoidal waveforms, these do not necessarily have an exact mathematical relationship with the rms. figures. They may also depend to some extentv on the instrument used.*

Table 6. **Vibration limits for tests I the manufacturers works**

Condition	Fan Application Category	Rigidly Mounted mm/s		Flexibly Mounted mm/s	
		peak	rms.	peak	rms.
Start-up	BV-1	14,0	10	15,2	11,2
	BV-2	7,6	5,6	12,7	9,0
	BV-3	6,4	4,5	8,8	6,3
	BV-4	4,1	2,8	6,4	4,5
	BV-5	2,5	1,8	4,1	2,8
Alarm	BV-1	15,2	10,6	19,1	14,0
	BV-2	12,7	9,0	19,1	14,0
	BV-3	10,2	7,1	16,5	11,8
	BV-4	6,4	4,5	10,2	7,1
	BV-5	5,7	4,0	7,6	5,6
Shutdown	BV-1	Note 1	Note 1	Note 1	Note 1
	BV-2	Note1	Note 1	Note 1	Note 1
	BV-3	12,7	9,0	17,8	12,5
	BV-4	10,2	7,1	15,2	11,2
	BV-5	7,6	5,6	10,5	7,1

Note 1. *Shutdown levels for fan in Fan Application Grades BV-1 and BV-2 must be established based on historical data*

Note 2. *The rms. values given in Table 5 are preferred. They are to a rounded R20 series as specified in ISO 10816-1. Peak values are widely used in North America. Being made up of a number of sinusoidal waveforms, these do not necessarily have an exact mathematical relationship with the rms. figures. They may also depend to some extent on the instrument used.*

Table 7. **Vibration limits for fan tests 'in-situ'**

15.   CONCLUSIONS

These two new standards are about to be published and will do much to regularise the measurement and specification of fan balance and vibration. As the first editions in this field they will no doubt rapidly be seen as in need of improvement. Nevertheless they meet a long felt want and give information, instruction and advice which is truly applicable to fans without the need for provisos or restriction. They should provide the basis for contractual agreement between specifiers, users and manufacturers and ensure that acceptable quality is achieved without specifying unattainable limits.

© 2000 With Author

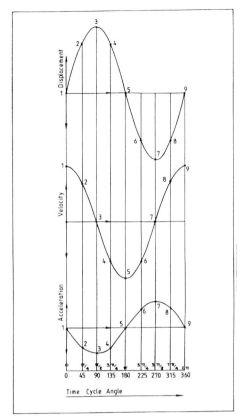

Fig. 1   Sinusoidal vibration

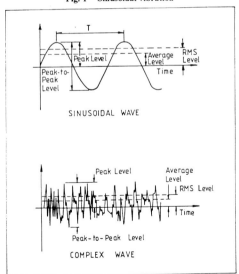

Fig. 2   Relationship of various vibration levels

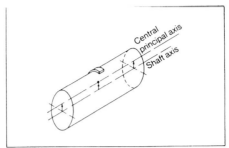

Fig. 3 Static unbalance

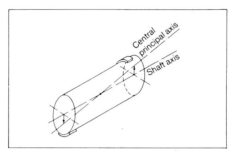

Fig. 4 Dynamic unbalance

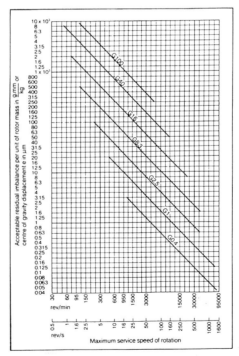

Fig. 5 BS5265 balancing quality grades

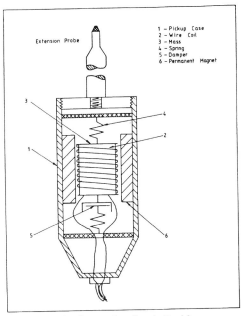

Fig. 6   Cross section of a velocity pickup

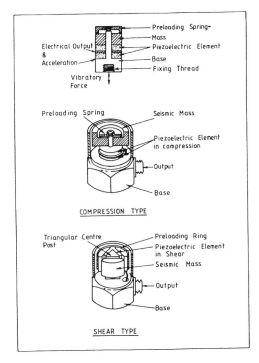

Fig. 7   Construction of accelerometers

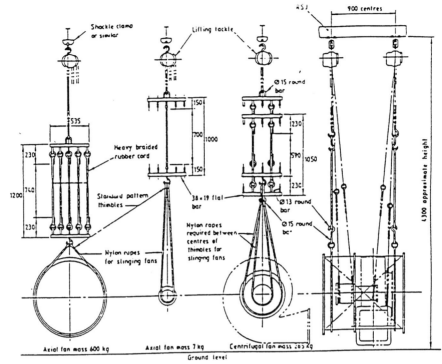

Fig. 8   Examples of resilient elastic rope mounted fans

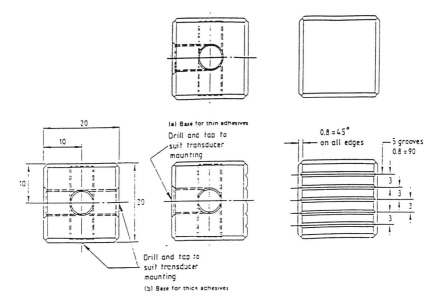

All linear dimensions are in millimetres.

NOTE 1   When mounting a velocity transducer a larger block will be required.

NOTE 2   When attaching a block to a tan by adhesive care should be taken to ensure the film is as thin as possible and the block is set true, i.e. vertically and horizontally. Adequate time should be allowed for the adhesive to cure. Typical adhesives are :

       epoxy resin
       plastic padding (hard)
       cyanoacrylate
       dimethacrylate urthane diol

Fig. 9   Typical mounting block for transducer

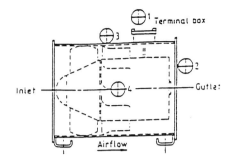

⊕ 1 and 2   Motor free end

⊕ 2   Axial measurements may not be possible when installed

⊕3 and 4   Motor supports fans end

Fig. 10 Recommended measuring postion for machinery health measurement of axial flow fans

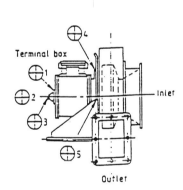

⊕ 1,2 and 3 Motor free end

⊕ 4 and 5   Motor fan end

Fig. 11 Recommended measuring position for machinery health measurement of centrifugal fans

ISO 13347 – the new standard for measuring noise by reverberant room, enveloping surface, and intensity methods

A GUEDEL
CETIAT, Villeurbanne, France

1) INTRODUCTION

The purpose of ISO 13347, which is being drafted by ISO/TC 117 - WG "Acoustics", is to define laboratory test methods for determining the airborne sound power level radiated in each 1/3 octave frequency band by industrial fans. This standard has been established by using existing national standards i.e. BS 848 Part 2, DIN 45 635 Part 38, NF S 31-021 and AMCA 300 and 320, which themselves are based on general acoustics standards drawn up under the responsibility of ISO/TC 43.

The text of the standard comprises four distinct parts, a general introduction and presentation of the three test methods adopted, i.e. the reverberant room, enveloping surface and intensity methods. Details of these four parts are given below.

2) PART 1: GENERAL OVERVIEW

Part 1 presents the objectives, field of application and general principles that are common to the different test methods.

The Standard 13347 describes the assembly and test conditions and the calculations for determining the sound power per 1/3 octave or octave band emitted by the free fan inlet and/or outlet as well as by the fan casing. In compliance with international terminology, used in particular in ISO 5801 [1], we can distinguish four installation categories for the fans:

- Type A: free inlet, free outlet
- Type B: free inlet, ducted outlet
- Type C: ducted inlet, free outlet
- Type D: ducted inlet, ducted outlet.

The level of inlet noise being generally different from the level of outlet noise and different from the casing noise, enables up to 12 different sound power levels to be distinguished on a single fan. Fan in-duct levels cannot be determined accurately from ISO 13347. It is much more preferable in this case to use ISO 5136 [2] which is adapted for that. For small unducted fans (type A), ISO 10302 [3], based on measurement with the aid of a mylar plenum, is recommended.

ISO 13347 does not apply to fans such as ceiling fans and table fans, for which IEC 60704-2-7 is best adapted, nor to jet fans, which can be tested in accordance with ISO 13350. ISO 13347 is first of all aimed at measuring fan noises on standardised test installations with engineering grade accuracy. This Standard does not deal with fan noise measurements on-site, even if we can use it to carry out such measurements, bearing in mind that the uncertainty of the result will be greater than on standardised installations.

As the noise level of a fan is highly dependent on its operating point, it is essential to be able to measure the aerodynamic performance of the fan, in accordance with ISO 5801, simultaneously with the acoustic performance. The specifications of the ducts and test assemblies described in this standard enable these measurements to be carried out simultaneously.

Whenever the fan is connected to an inlet and/or an outlet duct in which no sound pressure measurements are being taken, this duct, called a terminating duct, must be equipped with an anechoic termination in order to impose a well defined acoustic impedance and to thereby minimise the duct-end reflections that would distort the measurement of sound power radiated by the fan on the opposite side. The performance of the termination equipping a terminating duct does not need to be as good as that of a true anechoic termination mounted on a test duct (a duct in which sound pressure measurements are carried out in order to determine fan in-duct sound power levels), to the extent that a simplified anechoic termination can be used, the characteristics of which are described in ISO 13347.

3) PART 2: REVERBERANT ROOM METHOD

This part of the standard gives details of the method for measuring the sound power level radiated by the free inlet and/or outlet and by the casing of the fan in a reverberant room. It takes up the guiding principles of the North American standard AMCA 300 while relying on the general acoustics standard ISO 3743.

The environment of the fan to be tested must be sufficiently reverberant to satisfy a qualification procedure for the test room described in this part.

The method used is the method by comparison using a reference sound source, in accordance with the principle of ISO 3743-1. It is based on the fact that, in a given acoustic environment, the difference by frequency band between the sound power and the sound pressure is constant whatever the source of the noise.

Therefore, in the test room, a measurement is taken of the fan sound pressure level spectrum at the operating point required as well as the spectrum of the reference source and, knowing the sound power of the reference source, a calculation can be made of the sound power spectrum of the fan. In practice, the fan and reference source sound pressure spectra are measured in several discrete positions (or by continuous circular scanning) in the room and a spatial average of the measured spectra is carried out. In an "ideal" reverberant room the measurement on a single point would suffice, the sound pressure level being constant throughout the room.

Providing a suitable environment is available, this method is very practical to implement and certainly the one that, amongst all the different methods proposed in ISO 13347, leads to the shortest test period.

4) PART 3: ENVELOPING SURFACE METHOD

This method is based on ISO 3744 by using the assembly and test conditions of the Acoustics Standards for fans BS 848 Part 2, DIN 45635 part 38 and NF S 31-021. The test environment must be in compliance with the ISO 3744 specifications, i.e. being close to an acoustic environment of the free field type with one or several reflecting planes. In practice, measurements can be made both outside as well as inside the premises providing there is sufficient space around the fan and that the background noise is lower than the fan noise in each frequency band by at least 6 dB.

The sound power spectrum radiated by the fan is determined from the measurement of the sound pressure spectrum averaged at several points on a measuring surface surrounding the fan. Three different measuring surfaces are recommended in the standard: a rectangular parallelepiped (Figure 1), a large hemisphere resting on a reflecting surface (Figure 2) and a small hemisphere surrounding the fan air inlet (Figure 3).

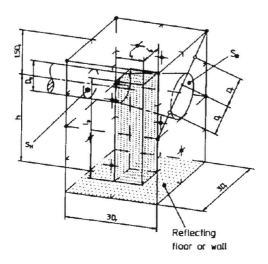

Figure 1: Enveloping surface method - Rectangular parallelepiped

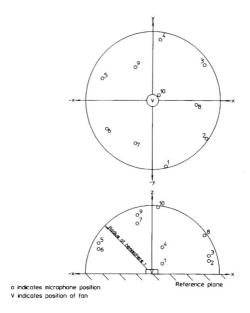

o indicates microphone position.
V indicates position of fan

Reference plane

Figure 2: Enveloping surface method - Large hemisphere

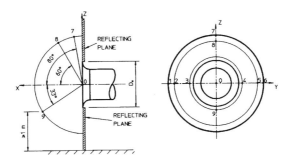

Figure 3: Enveloping surface method - Small hemisphere

In order to obtain the sound power spectrum of the fan, the test site must first of all have been qualified by determining the band environmental correction K_2 (in dB) as defined in ISO 3744. This correction is obtained using a reference sound source, mounted in place of the fan, based on the fact that, as already pointed out for the reverberant room method, in a given environment, the difference between the sound power and sound pressure levels is independent from the noise source.

The advantage of this method in comparison to the other two is that it does not need a specific acoustic environment or a special measurement device. It only requires the background noise level to be sufficiently low in relation to the fan noise.

5) PART 4: SOUND INTENSITY METHOD

This method is based on the principles of ISO 9614 and follows the specifications of AMCA 320 draft standard. It is based on the measurement of the sound intensity on a surface surrounding the fan, and necessitates having a sound intensity probe available with a properly adapted means of treatment. The test room must not be too reverberant, if not the method is inapplicable.

In order to qualify the system of measurement, the appropriateness of the site as well as that of the test operator, the sound measurements are conducted not only with the fan itself, but also with a reference sound source mounted in the test room in place of the fan. The difference between the sound power of the reference source measured by the intensity method and its power obtained from a calibration enables the correction of the power measurement carried out on the fan to be made, according to the same principle as that of the reverberant room method.

The advantage of this method in comparison to the two preceding methods is that it does not require a specific test room and that it is much less demanding in terms of background noise. On the other hand, it requires a properly adapted device and more thorough training for the test operator.

6) CONCLUSION

The objective of ISO 13347, which is in the process of being drawn up by the WG2 of ISO/TC 117, is to offer manufacturers and fan installers a certain number of standardised test methods enabling the sound power radiated by the free inlet or outlet and the casing of fans to be quantified. These methods are rigorous from a scientific point of view, because they are based on the principles of general acoustics standards drawn up by ISO/TC 43, and they are also adapted to industrial requirements because they aim in particular to allow for the simultaneous measurement of aerodynamic and acoustic fan performances. They also aim to simplify test assemblies and procedures as much as possible.

REFERENCES

[1]   ISO 5801: Industrial Fans - Performance testing using standardised airways (1997)

[2]   ISO 5136: Acoustics - Determination of sound power radiated into a duct by fans and other air-moving devices - In-duct method. (2000)

[3]   ISO 10302: Acoustics - Method for the measurement of air-borne noise emitted by small air-moving devices (1996)

The CEN standard for explosion proof fans to satisfy the Atex directive

H G D WITT
Witt and Sohn, Pinneberg, Germany

SYNOPSIS

This is a status report June 2000. Details may be updated up until October 2000. The coming standard is intended to specify design criteria for fans expected to function safely underground in coal mines or above ground in different industries. Ignition hazards due to gases, vapours, mist or dust will be considered.

FOREWORD

The Commission of the European Union, General Directorate III, has in March 1994 published directive 94/9/EC concerning equipment and protective systems intended for use in potentially explosive atmospheres, often called ATEX-100 Directive. This Directive is mandatory from 30.06.2003.

In view of this, The European Committee for Standardisation (CEN) has published the European Standard EN 1127-1 "Explosive atmospheres – Explosion prevention and protection – Part 1: Basic concepts and methodology".

This European Standard describes methods for the assessment of hazardous situations arising from explosive atmosphere and guidelines for the design and construction measures appropriate for the required safety. It is very general and applicable to any machinery to be used where explosions can occur due to combustible material emanating from the machine itself or other sources in the presence of air.

Presently a more specific standard is being developed exclusively dealing with the safety of the component "fan". It will give guidelines and rules, how fan manufacturers can comply with the safety requirements of the Directive.

1. INTRODUCTION

The new EN standard will contain general guide lines for the design of fans of category 1 G, 2 G, 3 G, 1 D, 2 D and 3 D to be used in explosion hazardous locations for gases, vapours, mist or mixtures thereof and for atmospheres carrying inflammable dust, G being an abbreviation for gas and D for dust. The above categories are identical with the well known definitions for zones 0, 1, 2, 20, 21 and 22.

It is the wish of the European Commission that the standard also should be applicable for mining fans category M1 which normally are energised in the presence of explosive atmosphere and category M2, fans normally not energised in the presence of explosive atmosphere. This meets with some problems and at present it is not finally decided if mining fans will be included in the general fan standard or if they will be subject to a different standard.

2. RISKS AND COUNTER MEASURES

Fans can represent a dangerous ignition source due to a number of different occurrences as:

a) from hot surfaces (e.g. from damaged bearings)

b) from mechanically generated sparks

c) hot spots as a result of friction or impact caused by contact between rotor and fixed components

d) from discharges of static electricity when using a non-conductive material

e) from foreign particles as metal, minerals or organic materials entering the fan

f) from rare malfunctions like broken blades

g) overheating of the gas

In addition, the prime mover, in most cases an electric motor, or other electric components, may present a hazard.

The requirements detailed in EN 1127-1 shall be deemed to have been met if a fan satisfies the specifications set down in the new document. The required level of protective or preventative design measures for fans depends on the likelihood of a hazardous explosive atmosphere occurring and the nature of the combustible component. Though all risks cannot be eliminated entirely, the following constructive measures aiming at avoiding ignition sources are specified here:

a) Larger air gaps compared with normal fans of equivalent size

b) Sturdy design

S723/007/2000

c) Positive locking of impellers, shafts and pulleys to prevent contact between rotating and static components

d) Proof motor and bearing lockings so no displacement can occur

e) Using metal pairings which give the lowest possible ignition hazard

f) Using non metallic material pairings only if these are proved to represent a low ignition hazard, from both elctrostatic discharges and by friction heating

g) Employing only electrical items of proved ignition preventing design

h) Mounting grids in order to prevent foreign particles to enter the fan

i) Safe enclosure of potentially explosive atmospheres

j) Labeling of explosion category outside the fan

k) Gas temperature control

The standard committee is trying to arrive at rules giving a very high degree of safety without prohibitive costs.

For zones 0 and 20 i.e. category 1 G and 1 D additional protective methods will have to be employed, in order to obtain a higher degree of protection. Fans for category 1 need furthermore a special accreditation as issued by one of a number of notified bodies.

3. RISK ASSESSMENT INSIDE AND OUTSIDE OF THE FAN

An explosion hazard may primarily exist inside and outside of the fan.

Fans which either convey an explosive atmosphere or are operated in hazardous areas or for which both conditions prevail, will have to be of explosion-proof design. According to the likelihood and duration of the existence of an explosive atmosphere the explosion category must be determined, which the fan must comply with. The inside and the outside of the fan must be considered separately, but not independently.

Fans, especially their shaft seals, elastic sleeves and flange connections can not be considered absolutely gas tight, nor will connected ducts be absolutely tight. Hazardous atmosphere may leak either from the inside of the fan into the fan environment or from the fan environment into the fan. Even in free space and normally clean air a hazardous environment around the fan caused by leakage has to be considered.

Table 1 gives the possible combinations of categories for fans for gases, vapours or mists as well as dusts.

Table 1 – Categories for fans (equipment group II)

Category outside the fan enclosure	category inside the fan enclosure				
	no category	category 3 gas and dust	category 2 gas and dust	category 1 gas	category 1 dust
no category					not allowed
category 3 gas and dust					not allowed
category 2 gas and dust					not allowed
category 1 gas					not allowed
category 1 dust	not allowed	not allowed	not allowed	not allowed	not allowed

The design criteria in each field of the matrix will be given and have to be complied with. Presently the committee thinks that category 1 dust can not be handled safely.

By convention the hazard difference between the inside and the outside of the fan is considered to be never more than 1 category step, so for instance an internal hazard 1G (G ~ gas) will automatically require that the fan outside is considered a category 2G hazard, even if the fan is placed in free field.

It has so far not been possible to define the size of the dangerous space surrounding a fan carrying an explosive medium, as it may depend on many parameters, but at least all integral fan components may be considered to be inside the zone, where the leakage hazard will have to be considered.

Some design criteria may be different for different categories while other criteria may be applicable to all categories. At this moment many details are still under discussion and no definite decisions can be reported here.

4. OPERATING TEMPERATURES

Explosion proof fans will normally be designed for a temperature range of – 20°C to + 40°C. If lower temperatures may occur, the manufacturer will have to choose alloys able to withstand these temperatures without becoming brittle. In this case the minimum permissible temperature must be indicated on the sign plate.

Higher gas temperatures are only permissible, if the fan is designed to guarantee a sufficient safety margin to the ignition temperature of the gas. This temperature is also to be indicated on the sign plate.

The safety margin must be chosen with due consideration of the increase in fan temperature due to compression work and fan losses. Closed dampers may not lead to an overheating of the gas.

5.1. Mechanical Design Criteria

Explosion proof fans must evidently be of a sturdy design. The committee has so far not defined details. The gap between rotating parts and housing is the most important safety feature of ignition risk reducing fans. Therefore much attention is paid to the fixation of the impeller to the shaft. At least in category 1 and category 2 fans a positive locking of the impeller will be prescribed. A number of recorded explosions have occurred because impellers have started to slide along the fan shaft, touching static parts of the fan. For the same reason sleeve bearings will only be allowed in small fans for category 3 applications, if at all.

Furthermore a minimum clearance between rotating components (e.g. impeller) and fixed components (e.g. housing) is defined, which at upstart must be at least 1 % of the possible contact diameter of the finished fan. This clearance shall not be less than 2 mm in axial direction and 2 mm in radial direction and need not be greater than 20 mm.

Shaft seals are not subject to this provision.

The clearance must be inspected before the fan is energized.

5.2. Fire Resistance

The materials used in the rotor and housing shall be able to withstand short-term exposure to flames. This requirement is met if the components when exposed to a propane Bunsen burner flame approx. 150 mm long for 30 seconds without additional air supply, are only partly destroyed without the onset of a self-sustaining combustion, as specified in EN 50018 (1994-08), or develop smoke in appreciable amounts, for instance 60 % maximum vision abstraction due to smoke according to BS 511.

5.3. Belt Drives

Belt drives represent a threefold risk. Electrostatic charging of the belts may lead to ignitions and if belts are too tight that may cause bearing overheating and eventually ignition. Slipping belts – f.i. during upstart or if the impeller is blocked - may create high temperatures due to friction. Therefore direct driven fans are preferred. When belt drives are used at all, the belts shall be made from antistatic material and shall be continuously monitored against slippage and breakage of the belt.

The electrical resistance of all drive belts shall satisfy the respective requirements of ISO 9563 or ISO 1813. Probably belts will only be allowed in small category 3 fans.

5.4. Materials and Components

In view of malfunctioning which can be expected to occur during normal operation, potential zones of contact between rotor and fixed components (with the exception of shaft seals) will have to be made of materials in which the risk of ignition through friction and friction-impact sparks is minimised or preferably eliminated. This requirement is deemed to be met if one of the acceptable material pairings described in EN 1127-1, sections 5.3.4 and 6.4.4, is used.

For fans, the following 3 pairings are preferred. Other materials could be employed, provided their suitability has been proved and documented.

a) Naval brass CuZn39Sn, DIN 17 660, Werkstoffnummer. 2.0530, equivalent to BS CZ112, combined with carbon - or stainless steel.

b) Leaded brass CuZn39Pb3, DIN 17 660, Werkstoffnummer. 2.0401, BS CZ121,combined with carbon - or stainless steel.

c) Silumin with Silumin, both having a maximum magnesium content of max. 4 %, e.g. DIN EN 573 T3 AISi10Mg, AIMg3 Mn0.4 or equivalent alloys described in other standards.

   Note: When silumin is used it must be ensured that no flying rust flakes can be deposited on surfaces that may come into contact with each other. Paints containing iron oxides or other materials reacting with aluminium are strictly prohibited.

d) Silumin according to c) combined with either naval brass, see a) or leaded brass, see b)

   Note: It must be ensured that no flying rust flakes can be deposited on surfaces that may come into contact with each other. Paints containing iron oxides are strictly prohibited.

e) Tin combined with carbon- or stainless steel.

   Note: The use of tin may be the best material when explosive dust with low ignition temperature is present, as tin will melt before dangerous hot spot temperatures are reached. On the other hand the low melting temperature may represent a risk for underlying materials to touch.

If linings are employed, they must have a minimum thickness in order to withstand impact or abrasion for some time. For a motorpower up to 10 kW this should be a minimum of 2 mm, for motor power between 10 and 30 kW this should be at least 3 mm, and up to 90 kW at least 4 mm. Linings shall be attached to the fan only using materials, which are compatible from a corrosion point of view.

The properties of plastics depends besides their chemical composition largely on the fillers being used. Therefore the use of plastic is not à priory permissible, as it may have low heat conductivity leading relatively easily to hot surfaces. If mechanical or electric sparks may occur depends also on the filler used and sparks cannot generally be excluded, without thorough testing of the specific composition and the fillers being used.

The use of steel alloys or cast iron combined with steel alloys or cast iron will probably not be permitted. Tests have shown that they may lead to hot surfaces or may cause mechanical sparks. There is some violent opposition against this proposed rule from some manufacturers, however, who wish to allow iron or steel for low motor powers in category 3 G fans.

5.5. Electrical Components

Electric motors for all explosion categories must have a sufficient power reserve, so the motor is not overloaded if the system resistance is different from the value calculated or if a damper can produce a higher resistance. Special attention must be paid to impellers with forewards bend blades.

The Electrical components themselves are outside the scope of the fan standard. A number of EN standards for electric motors and other components exist and have to be observed.

5.6. Vibration Characteristics

Vibration limitation will receive much attention in the standard as vibrations may lead to fatigue and premature breakdowns, which in turn might lead to ignitions. Increased vibrations may also be an indication that the fan is damaged.

At startup the operating speed must differ by at least 20 % from the actual critical shaft speed. In case of over-critical operation the drive must ensure rapid acceleration through the critical rotational speed.

The impeller shall be properly balanced (quality grade G 6.3 or better in accordance with ISO 1940-1). If the fan has a belt drive the pulleys shall also be balanced to the same quality grade G 6.3 or better.

6. MAINTENANCE

A high level of maintenance will be required.

© 2000 Hans Witt

S723/008/2000

The other fan standards – dimensions, vocabulary, and definitions of categories, tolerances, methods of conversion and technical data presentation, and performance testing of jet fans

W WOODS BALLARD
Consultant, Ipswich, UK

1. INTRODUCTION

The above, 'Other Standards', will become, respectively, Parts 4 (ISO13351), Part 8 (13349), Part 9 (ISO13348) and Part 10 (ISO13350) of BS848. Those that are underlined are already published. The detailed status of the BS848 fan standards, and their corresponding ISO references, is given in the table included in the binder. Like all the fan standards, those in this short paper have benefited from the technical expertise and good practice available from national and international trade associations, relevant national research organisations and national standards bodies. To evolve the new fan standards, all this information and expertise, together with relevant new legislation, has been sifted and refined many times over by members of the ISO Technical Committee for Industrial Fans - ISO TC/117.

2. DIMENSIONS – ISO 13351

2.1 History

In 1981, with the title 'General Purpose Industrial Fans - Circular Flanges', ISO6580 was published. Largely based on the experience of one manufacturer, this initial series of metric flange details, with only minor changes, was adopted for the class of General Purpose Fans. The publication had a good deal of input and support from Eurovent. However, when it was later suggested that a range of standardised rectangular flanges, to complement ISO6580, would be of help to the fan industry, support was not forthcoming through Eurovent. Subsequently, and as part of the work programme of ISO/117, the need for a series of standardised rectangular flanges was accepted, along with the benefits of specifying the details of an additional range of heavy duty circular flanges. Clearly, a definition of 'Fan Size Designation' had to become part of the standard. Thus, BS, 848: Part 4: 1997 (ISO 13351: 1996), whilst allowing scope for individual manufacturers' best practice, contains the characteristics for two series of standardised circular flanges and a range of rectangular flanges in addition to the definition of Fan Size Designation.

2.2 Designation of fan size

The fan size is designated by the 'nominal' impeller tip diameter. This being the impeller tip diameter on which the design of the fan is based.

2.3 Circular flange characteristics

Unlike ISO 6580, which becomes obsolete in 2002, ISO 13351 contains two series of circular flanges. One for 'general purpose industrial fans' and a second for 'heavy duty fans' such as may be used for marine or heavy industry applications. Whilst there is no need to comment specifically on the 'heavy' flanges, there were many changes between ISO 6580 and ISO 13351 with regard to the 'general purpose' range. The revisions were the outcome of a close examination of both good engineering practice and practical aspects such as accessibility. The major differences are summarised in the table below. However, it must be emphasised that none of the flange details, in the R20 range covering 100 to 2000 mm, are identical and manufacturers who have embraced ISO6580 will have to incur considerable costs in making the change to the new BS 848: Part 4: 1997 (ISO13351)

Feature/ISO	13351(6580)				
Fan Size 'D'	280	450	1000	1400	2000
P.C.D. 'd_1'	332(320)	497(500)	1067(1070)	1475(1470)	2073(2080)
Bolt Dia. 'd_3'	M10(M8)	M10(M10)	M12(M12)	M12(M16)	M16(M16)
Hole No's. 'N'	4(8)	8(12)	24(16)	20(32)	24(40)
Arc Length 'l'	130(251)	130(196)	140(210)	145(231)	163(272)

2.4 Rectangular flange characteristics

As with the circular flange system, the rectangular flange standard provides a maximum of freedom of choice whilst maintaining the ISO Preferred Number Series for the basic dimensions. The flange details allow for the option of corner holes and cater for four alternative flange fixing methods. For aspect ratios significantly different from 1, which can result in undesirable differences in flange details, guidance is given on the procedure to follow to achieve an acceptable result.

3. VOCABULARY AND DEFINITIONS OF CATEGORIES – ISO 13349

3.1 Introduction

The need for an International Standard has been evident for some time. A number of national documents, focussed on national practice, have been in existence for a while. In many instances these either described features in a different way or did not cover the engineering solutions developed in another market. Wherever possible, in the interests of international comprehension, this Standard is in agreement with similar documents produced by EUROVENT, AMCA (USA), VDMA (Germany), AFNOR (France) and UNI (Italy). These have, however, been amplified or added to where the need was apparent. Use of the Standard should lead to a consistent approach in specifying fan equipment and manufacturers, consultants, contractors and users are encouraged to use it. To assist in this process a sample definition for a centrifugal and and an axial fan is included in the Standard and is repeated at the end of this paper.

The more important sections of this Standard are:

3	Definitions
5.2 - 5.3	Fan categories
5.4	Drive arrangements
5.5	Inlet and outlet conditions
5.6	Method of fan control
5.7	Designation of direction of rotation. Position of parts of the fan
5.8	Characteristic dimensions and component parts

3.2   Fan categories

As stated in Section 5.1 of this Standard, fans may be categorized according to:

a)   suitability for the fan pressure;

b)   suitability of construction (including features required for smoke ventilation, gas tightness and ignition protection);

c)   driving arrangement;

d)   inlet and outlet conditions;

e)   method of fan control;

f)   rotation and position of parts;

g)   characteristic dimensions and component parts.

Section 5.2 concerns 'suitability for the fan pressure' and Table 1 - Categorisation of fan according to level of work per unit mass - shows the codes and categories to be used whilst the text indicates where compressibility must be considered. 'Suitability of construction' is dealt with in Section 5.3. Table 2 - Categorisation according to method of construction - shows typical casing features and methods of construction for a number of common fan applications. Table 3 - Recommended categorisation of smoke-ventilating fans - is repeated below:

Temperature Category	Coding	Max air/gas temp °C	Minimum operating time h
A	HT/150/5.0	150	5.0
B	HT/200/2.0	200	2.0
C	HT/250/1.0	250	1.0
D	HT/300/1.0	300	1.0
E	HT/400/2.0	400	2.0
F	HT/600/1.5	600	1.5

Table 4 of Section 5.3 has the title, 'Categorisation of gas tight fans - Leakage as a function of pressure'. This table has been put together as an amalgum of practice and experience in the air-conditioning, processing and nuclear industries. The notes in the text indicate which 'Leakage category' is appropriate to which application. Spark resistant construction is also addressed in Section 5.3 under Table 5 - Categorisation of ignition protected fans. The table, which must be used in conjunction with Section 5.3.5 of the standard, is based on industry practice. It does not cover all fan construction materials and for further information reference should be made to the EUROVENT document "Ignition -protected fan". Table 5 is reproduced below. However, this may well be overtaken by the ATEX Directive and a CEN Standard on Explosion Proof Fans.

Type	Construction
A	All parts of the fan in contact with the air or gas being handled shall be made of metals that do not produce sparks or hot spots when striking or rubbing against each other, which may ignite the gases, liquid or dust that may be present. Steps must be taken to ensure that the impeller bearings and shaft are adequately attached and/or restrained to prevent lateral or axial shift in these components.
B	The fan shall be designed so that the parts of the fan that are most likely to touch in the case of fault are made of metals that do not produce sparks or hot spots when striking or rubbing against each other. This could in an axial fan be obtained by e.g. aluminium impeller blades and an aluminium lining to the casing, other parts such as the hub, shaft etc. may be a ferrous material. Steps shall also be taken to ensure that the impeller, bearings and shaft are adequately attached and/or restrained to prevent a lateral or axial shift in these components.
C	The fan shall be so constructed that a shift of the impeller or shaft will not permit two parts of the fan to rub or strike, which might produce sparks.
NOTE Where low risk materials are not used (e.g. some auxiliary mines fans) all clearances between rotating and static parts shall be not less than 5 mm.	

3.3   Drive arrangements
The six most commonly used drive arrangements are;
a)   Direct drive from the shaft of the motor or prime mover
b)   Direct drive through an inline coupling
c)   Drive through an in-line, slipping coupling
d)   Drive through a gearbox
e)   Belt drive
f)   Direct drive with inset motor
Tables 6 and 7 show, for centrifugal and axial units respectively, the classification according to drive arrangement. Many of these are the same as those in use by AMCA and UNI. There are 19 different arrangements for centrifugal fans and 8 for axial fans. Where appropriate, the same arrangement number is used for both centrifugal and axial fans.

3.4   Inlet and outlet conditions
The direction or condition of the flow into or out of the fan may be modified by the addition of ancillaries connected directly to the fan. These are identified, diagrammatically, in Table 8 in Section 5.5. Eight inlet/outlet ancillary options are shown for three basic fan types: axial flow, single inlet centrifugal and double inlet centrifugal.

3.5   Method of fan control
Section 5.6 describes four basic methods of controlling fan output:
a)   Variable speed control
b)   Damper control
c)   Inlet vane control
d)   Blade pitch control either whilst the fan is operating - variable pitch - or when stationary - adjustable pitch

3.6   Designation of direction of rotation and position of parts of the fan assembly
Designation of these aspects is contained in Section 5.7. Essentially the same as EUROVENT publication 1.1, they are clearly illustrated in Figs 15 to 23, whilst appropriate examples are given in the text.

3.7 Characteristic dimensions and component parts

The key for this part of the standard, Section 5.8, is Table 9. Table 9, with the title: "Index illustration of fans", is reproduced below:

Ref.	Fan type	Features
Aa	Centrifugal	Backward curved - indirect drive
Ab		Forward curved - direct drive
Ac		Paddle blades - indirect drive
Ad		Vane controlled - coupled drive
Ae		Double inlet
Af		Multistage
Ag		Two stage with duct connection (duplex)
Ba	Axial flow	Long casing - guide vanes - direct drive
Bb		Short casing - direct drive
Bc		Indirect drive
Bd		Shielded motor (bifurcated) - direct drive
Be		Multistage - indirect drive
Bf		Propellor fan
Ca	Mixed-flow	Direct drive
Da	Cross-flow	Direct drive

The 'References', Aa to Da, are illustrated in figures 24 to 27 as chosen examples to show component parts of fans. Their descriptions are given in Table 10.

3.8 Example use of standard

Coupling-drive heavy-duty centrifugal fan with impeller between bearings and designed for a duty of 38 m^3/s against a fan pressure of 6.3 kPa.

Casing to be suitable for supporting associated ducting. Side boxed inlet with vane control and diffuser on the outlet, terminating in a flange to match client's ducting. Casing to be fitted with an inspection door and to be suitable for handling radioactive fumes without detectable leakage.

Type of installation	Type "D"	3.4, figure 1
Type of fan as a function of its role	Ducted	3.5
Type of fan according to fluid path	Centrifugal	3.6
Suitability for pressure	High pressure, Category M/6	5.2, table 1
Casing construction	Category 3	5.3.1, table 2
Temperature category	Gastight fan, Category G	5.3.3, table 3
Drive arrangement	Coaxial coupling, Arrangement 7	5.3.4, table 6
Inlet/outlet conditions	SD	5.5, table 8
Method of fan control	Vane control	5.6
Component parts	Outlet RD45	5.7, figure 19
	Inspection door RD315	5.7, figure 20
	Inlet box RD0	5.7, figure 20
Motor position	In-line	5.7, figure 20

4 TOLERANCES, METHODS OF CONVERSION AND TECHNICAL DATA PRESENTATION – ISO WD13348

4.1 Introduction

Though some standards and documents cover specific aspects dealt with in this standard, no one document covers them all. ISO/TC117 has recognised the need to improve both the precision of technical aspects of contracts, where fan performance is concerned, and the accuracy and consistency of data published in technical catalogues. To this end the document includes methods of performance conversion, tolerance on technical parameters and recommendations on data presentation.

Sections 1 to 4 cover the usual aspects of Scope, References, Definitions and Symbols and Units. The essential part of the standard starts in Section 5.

4.2 Information

So that a potential supplier has a reasonably complete idea of the purchaser's needs, Section 5 details some 9 operational requirements which may be essential to the specification of the equipment best able to satisfy the customer. In addition, where appropriate, the tolerance grade may be specified and the customers preference for fan type construction materials, arrangement and drive (BS 848 : Part 8) stated.

In many instances where a series produced fan is involved the supplier can refer to published catalogue information and data. However, where a special purpose fan design and installation is involved a considerable amount of information may have to be generated and forwarded to the prospective customer. Section 6 gives recommendations regarding the scope of operational and engineering data which it may be necessary to provide.

4.3 Tolerances

Tolerance grades are dealt with in detail in Section 7 where the standard acknowledges that there are two different sources of uncertainty influencing performance tests. Manufacturing processes have inherent dimensional tolerances which can affect such important features as running clearances and design constraints, often generated by cost pressures, can result in. losses being less predictable on the smaller sizes. The second source of uncertainty is caused by the limitations in the measurement accuracy of the instruments being used for a performance test and the human error associated with reading them. A table – Guide to typical Fan tolerance grades, is reproduced below.

Certified Ratings programmes, outlined in "Informative Annex D" also touch on the subject of fan tolerances. These programmes are generally applicable to geometrically similar, series produced fans.

Four grades of tolerance magnitude are given in the table for the fan characteristic, power and sound level. These grades are applicable to contractual agreements between a manufacturer and the chosen supplier and where achievement of a specific performance is one of the requirements. The operating point is assumed to lie within + 10% of the point of optimum efficiency or within the optimum operating range as specified by the supplier. Section 9.2 gives guidance on tolerance grades for cases where the fan has to operate away from the optimum range specified by the supplier. Implementation of the tolerances is shown in Fig. 9.1.

When a customer specification calls for measurements to be carried out to verify compliance with the technical requirements, it shall be agreed as to how this should be done. The options are:

a)   performance testing to ISO5801, ISO13347 and ISO14694

b)   performance testing using a geometrically similar model, according to the standards detailed above, and converting using the Fan Laws described in this standard

c)   performance testing on site to ISO5802 noting that this introduces an additional uncertainty and also that, where inlet and outlet connections to the fan are not straight, then an unmeasurable System Effect Factor may need to be added.

Figs 9.1 and 9.2 (not reproduced in this paper) show the application of the tolerance zone to the specified parameter whilst reference to BS 848: Part 1 (ISO5801) gives guidance on the measurement uncertainty and the limits of error in measuring equipment, in Section 17.

Table - Guide to typical Fan tolerance grades

Tolerance Grade	Criteria		
	Application	Manufacturing processes and materials used for major aerodynamic components	Approximate Minimum Power, kW
AN1	Mining (e.g. main fan), process engineering, power stations (e.g. exhaust fan) wind tunnels, tunnels, etc.	Machined in some places, cast (high accuracy)	>500
AN2	Mining, power stations, wind tunnels, tunnels, process engineering, air conditioning	Sheet or plastic material, partly machined, cast (high accuracy)	>50
AN3	Process engineering, air conditioning, industrial fans, power station fans, industrial fans for harsh (abrasive or corrosive) conditions	Sheet material, cast (medium to low accuracy), special surface protection (e.g. hot-dip galvanising), moulded plastics	>10
AN4	Process engineering, ships exhausts, agriculture, small fans, industrial fans for harsh (abrasive or corrosive) conditions	Sheet material, special surface protection (e.g. rubber coating), moulded or extruded plastics	-
ANC	Series produced fans whose performance is pre-published in a catalogue and is accredited by a third party certification scheme	Sheet metal, cast metal, moulded plastic etc.	>0.5

Table - Manufacturing Tolerance Grades

Parameter	Limit deviation, t, for tolerance grades			
	AN1	AN2	AN3	AN4
Volume flow rate, q_v	± 1%	± 2.5%	± 5%	± 10%
Fan Pressure, p_F	± 1%	± 2.5%	± 5%	± 10%
Power, P_r	+ 2%	+ 3%	+ 8%	+ 16%
Efficiency, η	- 1%	- 2%	- 5%	- 12%
A-weighted sound power level, L_{wA}	+ 2 dB	+ 3 dB	+ 4 dB	+ 6 dB

Subsequent to the circulation of the 6th Working Draft, from which these extracts were taken, it has been agreed that two separate chapters will be introduced covering the general area of tolerances, conversion and acceptance criteria. Series produced fans, ANC tolerance grade, where a published technical catalogue is available, will be based closely on existing AMCA Certified Ratings procedures. More specialised fan equipment, tolerance grades AN1-4, are likely to be the subject of contract acceptance tests based on DIN procedures which form a large part of the current Working Draft.

4.4   Conversion of Air Performance Test Data

As well as giving information, by reference to ISO 5801 (BS 848: Part 1), on Similarity and the Fan Laws and Conversion Rules, Section 10 gives guidance on conversion rules and limits for Series Produced Fans where the effect of compressibility is assumed to be negligible. Extrapolation and interpolation of test data is also addressed.

4.5   Conversion of Sound Power Test Data

Conditions and limits to the application of conversion rules to total sound power level are given at the beginning of Section 11. This is followed by a guide to "Generalised Methods for Sound Power Level Prediction" where a procedure is outlined for the prediction of octave and one third octave levels. The method, which originates from a HEVAC document on Fan Technical Data Presentation, only applies to aerodynamically generated sources of noise and geometrically similar fans operating at the same flow coefficient. At the end of the section, tables are given to show the octave and one third octave centre frequencies and the 'A' weighting corrections for each octave band.

4.6   Technical Data Presentation

The purpose of Section 12 of the standard is to recommend a clear unambiguous and uniform presentation of fan technical data in publications. The section is intended for use by manufacturers of series produced fans and will generally apply to fans where the flow can be considered as incompressible. Included is typical 'outline drawing' data, the recommended 'chart' parameters along with possible additional sound, aerodynamic, mechanical and electrical data. The section finishes with some guidance on possible ways of showing performance data for installation categories other than that for which the fan was tested. Such corrections should be based on test and an indication of their order of accuracy given.

5 PERFORMANCE TESTING OF JET FANS – ISO 13350

5.1 Introduction
The use of the so-called Jet fan to assist in controlling the quality of air in vehicle and train tunnels has become increasingly popular. The longitudinal method of ventilation can show advantages in initial cost compared to alternative systems, and smoke control in emergency conditions can be readily provided. Until now there has been no published national or international standard for the performance testing of Jet fans. This standard, which covers laboratory testing, deals with the determination of those performance criteria essential to the correct application of Jet fans.

5.2 Definitions
Two of the more important definitions used in the determination of Jet fan performance are: 'effective fan outlet area' and 'effective fan outlet velocity'.

A_{eff}

In the particular case of a Jet fan, outlet area with deductions for motors, fairings or other obstructions – illustrated in Fig. 1.

v_{eff}

calculated from the thrust, the inlet density and the effective fan outlet area as detailed below:

$$v_{eff} = \left| \frac{T_m}{A_{eff} . \rho_a} \right|^{\frac{1}{2}}$$

The effective fan outlet velocity is used to calculate the correction factor 'k' on the thrust due to the tunnel mainstream velocity v_t, where:

$$k = \frac{v_{eff} - v_t}{v_{eff}}$$

5.3 Characteristics to be measured
Thrust – to meet specification

Input power – in order to be able to calculate operating costs

Sound level – to meet specification

Vibration velocity – safety and reliability

Volume flow rate – only if required for contractual reasons. V_{eff} is used to evaluate the optimum number, size and spacing of Jet fans in a tunnel.

5.4 Determination of thrust
There are two basic configurations acceptable for the determination of thrust: suspended configuration, of which there are three and supported configuration for which there are two methods. The first methods require that the suspension elements be kept precisely vertical and parallel with a vertical plane passing through the fan axis. The second method requires accurate construction and levelling of the support assembly. In either case force can be determined by the use of calibrated weights, spring balance or force transducer.

Whilst Fig.7 shows the dimensions of a suitable thrust measuring enclosure, aimed at keeping the general velocity ≤ 0.3 m/s, the test procedure states that the thrust measurement should only be recorded when the power and thrust readings have stabilised, or at least 10 min after start-up.

5.5   Determination of sound level

Sound levels are measured by the semi-reverberant method. Essentially a practical method, apart from sound measuring equipment, minimal facilities are required: a suitable enclosure, as shown on Fig.8, and a calibrated sound source.

As the Jet fan has only one operating point – at zero resistance – there are no complications which could arise from noise generated by the "loading means". Similarly, since only open inlet or open outlet sound levels are required, anechoic terminators are unnecessary. Because the test configuration enables all the noise; inlet, outlet, casing radiation etc, to be measured, the same situation as when the fan is installed in a tunnel is represented.

5.6   Determination of vibration velocity

Owing to the single operating point and the axial symmetry of the Jet fan, measurement of vibration velocity is a simple procedure. Only the vertical vibration velocity at the upstream and downstream attachment points needs to be recorded. A typical test arrangement is shown on Fig. 9 whilst maximum acceptable vibration velocities are given in Table 2.

5.7   Tolerances and conversion rules

Tolerances, intended to take into account measurement uncertainty and manufacturing variations, are given in Table 3. However, it is important to read the notes attached to Table 3 particularly where it concerns motor efficiency – Notes 5 and 6.

When direct tests are not available, and subject to agreement with the client, the conversion rules in Annex C can be used. However, it must be appreciated that these rules apply to geometrically similar Jet fans as outlined in Section 12.2. As true geometric similarity is unlikely to be achieved between the test fan and another fan, the applicability of the rules is limited to $\pm$ one R20 step in size and the test speed $\times$ 1.3 or test speed $\div$ 1.3.